Quality Control in
Analytical Chemistry

CHEMICAL ANALYSIS

A SERIES OF MONOGRAPHS ON
ANALYTICAL CHEMISTRY AND ITS APPLICATIONS

VOLUME 60

A WILEY-INTERSCIENCE PUBLICATION

JOHN WILEY & SONS
New York / Chichester / Brisbane / Toronto

Quality Control in Analytical Chemistry

G. KATEMAN

F. W. PIJPERS

Department of Analytical Chemistry
Catholic University of Nijmegen
The Netherlands

A WILEY-INTERSCIENCE PUBLICATION

JOHN WILEY & SONS

New York   Chichester   Brisbane   Toronto

Library of Congress Cataloging in Publication Data:

Kateman, G
   Quality control in analytical chemistry.

   (Chemical analysis; v. 60 ISSN 0069-2883)
   "A Wiley-Interscience publication."
   Includes index.
   1. Chemistry, Analytic—Quality control.
I. Pijpers, F. W., joint author. II. Title.
III. Series.

QD75.2.K37   543   80-23146
ISBN 0-471-46020-6

Printed in the United States of America

10 9 8 7 6 5 4 3 2

PREFACE

As a manager of a control laboratory of a large chemical plant, one of the authors of this book was continuously confronted with questions on quality, the quality of our product as compared with that of our competitors and the quality of the products of our suppliers. However, the questions to answer were always the following:

1. What is the quality of our (analytical) work?
2. Do we provide the people in the factory and the marketing people with the right answers to their questions?
3. Do they comprehend, trust, and use our data?

In our opinion every analytical chemist should be asked these questions, or he should ask them himself, and be prepared to answer them.

To answer these questions and to improve the quality of analytical chemistry requires a wide view of how quality can be affected. This is the reason that we elaborate on so many different techniques in our curriculum for analytical chemists at the Catholic University of Nijmegen, that we do research in some unusual techniques such as "queueing," and that we wrote this book.

Quality can be defined, and is defined in this book, as "the (numerical) value of a set of desired properties of an object (i.e., chemical analysis)." Thus quality is not an absolute property but depends on the definition of "desired."

In analytical chemistry quality can be associated with accuracy and/or reproducibility. Another quality criterion can be speed, cost, or information flow per dollar. Chemical analysis is always associated with an object to be analyzed. Furthermore, the analysis always has a purpose: to control, to describe, or to monitor the object. So a way of quantifying quality can be to answer the question; how well is the subject of analysis described in view of the applicability of the results?

In this book much attention is paid to the description of the work of the analyst—not a description of the various analytical chemical hardware and methods, but an explanation of how the result of the analysis is affected by the various stages of the analytical procedure: sampling, analysis, data handling, and, when aiming for a specific goal, control, description, or monitoring. When the reader has learned how quality in

v

chemical analysis may be affected in a great variety of practical situations, it will be quite easy for him to control the quality by optimizing a limited number of selected parameters.

Of course the quality level of analytical chemical work is determined by factors such as the expertise of the analyst, the quality of the description of the analytical procedure, and the performance of the analytical instruments. However, to attain a high standard, to convince oneself and others that the work cannot be done better, and to make clear in quantitative terms how good the quality of the work is, one needs methods to measure and control the quality of analytical chemical work.

G. KATEMAN

F. W. PIJPERS

Nijmegen
January 1981

ACKNOWLEDGMENTS

This book is the product of many discussions with, and is based on the lectures and papers of, quite a few people. We wish to express our appreciation to all those colleagues who provided material. Special thanks are due to Auke Dijkstra and Jaap Bruins Slot of the chemometrics group of the State University of Utrecht and to Charles Limonard, Piet Müskens, Henny Poulisse, and Bernard Vandeginste of our theoretical analytical chemistry group of the Catholic University of Nijmegen. We are much indebted to Lea Pöttgens-Debets, who played an important role in the compilation and writing of large parts of the original Dutch text. The majority of the figures was drawn skillfully by Peter van der Wiel, many of them after ideas of Lea Pöttgens-Debets. Enid and Dick Richmond had a heavy task in rephrasing our often rather cryptic language into more readable English. Typing a manuscript, and especially one with many equations, is a tedious task. One of us had convinced himself of that by typing large parts, assisted by his son Theun-Gerrit Kateman, until Annie Jo Kiem Tioe and Wilma Philipse finished that job.

G. K.

F. W. P.

CONTENTS

INTRODUCTION

Quality control in chemical analysis is an old concept. As long as there has been chemical analysis, there has been the need to control the quality of its performance. This book describes techniques that can be applied to improve the quality of analytical methods, sampling, and data handling. Often the results of these techniques can be used as a criterion for quality. The management of a laboratory has a need to estimate the quality of its efforts and, if this quality doesn't meet the standards, to improve it.

This statement leads to several questions. Why do laboratories want to improve their quality level and what happens if they don't? What quality parameters can be improved and which one should be improved first? How can quality be improved or controlled and why is it so difficult to keep quality on a steady, high level? Quality can nearly always be improved, but at the expense of much ingenuity, work, and money. Sometimes the efforts are required to reach a goal; sometimes reasons such as survival of the fittest are at stake. It would be desirable if some criterion could be applied that allows the optimization of quality; that is, the sum of the yield of the object of analysis and the negative yield (cost) of the analysis itself should be maximal. This yield may be expressed in terms of money, but, though money is a good and dependable yardstick, it is not the only one. Efforts in human ingenuity, although these are difficult to measure, could be the parameter of optimization. The amount of manpower or the level of sophistication of the instruments required can serve as a measure of ingenuity.

At this time there is no universal standard in operation. Until recently, accuracy and precision were the only quality parameters in use. These could not be optimized since it was impossible to estimate the yield and the cost corresponding to the various levels of accuracy, for example.

The gradual change in attitude toward the concept of quality control in chemical analysis has only a short history, though the first applications of quality control date back to the beginning of the century. In 1908, Gosset (GO 08), an amateur of statistics and a professional chemist, published a paper on the error of a mean under his pen name, "Student." He introduced a parameter known since then as "Student's-t." It allows analytical chemists (and others) to compare quantitatively the results of two items, such as methods of analysis or samples. Here the quality parameter in use is the level of probability that the two means match. Another statistical parameter that could be used is the standard deviation. This quality parameter combined with Fisher's "F-test" indicates the probability that the scatter in the results of a particular experiment is better or worse than the scatter found in a very reproducible standard series of analytical results. Although these quality parameters were known, they were not in common use among analytical chemists. In 1928 Baule and Benedetti-Pichler (BA 28) published a paper on the minimum size of samples of inhomogeneous lots, in which they indicated the quality parameters for sampling.

The first major impact on quality control in analytical chemistry was caused by the introduction of the control chart, familiarized by Shewhart in 1931 (SH 31). Such a chart allows the continuous supervision of the quality parameters accuracy and precision in an analytical laboratory. It enables corrective measures when a control line is crossed. A serious drawback of this method is the position of the control lines that is arbitrarily chosen.

After 1940 a host of publications appeared on the application of statistical techniques in analytical chemistry. However, most of these were directed to the application of statistics in the validation of analytical results. Its main purpose is the determination of the probability that certain effects in objects or processes can be detected by analytical chemical results. In analytical chemistry the application of statistical techniques for the control of analytical procedures was restricted to the comparison of analytical procedures, without establishing more quality parameters for analytical methods than standard deviation and variance. See, for example, AL 69, BE 75, DA 72, DI 69, DU 74, NA 63, FI 55, and YO 51. In 1963 van der Grinten (GR 63, GR 65, GR 66) published some simplified equations that were derived from the rules of Wiener on control theory. This allowed the quantitative evaluation of the quality of measurements with respect to the object that was measured. In the years that followed van der Grinten together with Kateman explored the possibilities of the quality parameter "measurability" at the chemical works of DSM in the Netherlands. To each component in a chemical process that was sampled and analyzed one could attribute a quality number when some parameters of the process (time constant and standard deviation of the uncontrolled process) were known. In 1971 Leemans (LE 71) published some of the results in the analytical literature. It became clear that, apart from accuracy and reproducibility, also the speed of analysis, the frequency of sampling, and their merit for the application of the results could be quantified and optimized.

These developments were not the only ones that caused a new view of quality control in chemical analysis. Around the year 1970 a discussion of the necessity of analytical chemistry arose—mainly in the American analytical literature—that culminated in the rhetorical question, "Analytical chemistry, a fading discipline?" (FI 70). The discussion was pursued in Europe and resulted in a series of definitions. These were formulated by Kateman and Dijkstra (KA 79) within three categories of definitions:

1. Analytical chemistry* produces information by application of available analytical procedures in order to characterize matter by its chemical composition.
2. Analytical chemistry* studies the process of gathering information by using principles of several disciplines in order to characterize matter or systems.
3. Analytical chemistry* produces strategies for obtaining information by the optimal use of available procedures in order to characterize matter or systems.

In a way these categories correspond to three distinct nonhierarchical levels in analytical chemistry. The first level refers to the actual production of analytical results. For this production instruments, procedures, and skilled personnel are required. The second level covers the research and development of analytical procedures. The third can be considered as an organizational level. It comprises the interaction between men and machines, communication as well as the optimal use of the tools available for producing information. In practice, of course, these levels are interwoven, but the division seemed to be indispensable for a new view on quality control.

In Europe the Arbeitskreis "Automation in der Analyse" (AR 73) formulated the view that analytical chemistry should be considered as a "black box," a system with known input and output and rules interconnecting these. According to this point of view, operations research, the science that studies this kind of system, could be applied to analytical chemistry. Massart (MA 75, MA 77) presented the first survey of research using this new approach. According to this approach, the application of operations research techniques to analytical chemistry indicates some new quality parameters:

*   The degree of optimization of a procedure, if quantitative optimization methods are used.

*Or analytical chemists.

- Information as a measure of quality of a qualitative method.
- Waiting time as a measure of planning and routing in analytical chemical laboratories.

A third phenomenon, parallel to the use of control theoretical concepts and operations research, was the introduction in analytical chemistry of integrated systems of statistics and correlation analysis, known as pattern recognition (KO 72). This technique allows, for instance, the selection of those analytical results that are indispensable and sufficient for describing a certain phenomenon. The quality of analytical chemistry, expressed in relevance of measured data, seems possible now. Moreover, it might be possible to select an analytical method that, given the properties of the problem under investigation, will give an optimal solution of the problem (VA 77).

Although the methods mentioned here sometimes are grouped under the heading "chemometrics" (KO 75), it does not seem justifiable to use this term for this book, since we confine ourselves to the problem of quality control. This means that many of the techniques that are covered by chemometrics can be used, but only with one purpose in mind: to help in the validation of analytical actions, procedures, and organizations, that is, to establish quality in the broadest sense of the word.

Wilson (WI 79) described an approach for achieving comparable analytical results that involves the sequential completion of a number of closely linked stages, beginning with the establishment of a working group and ending with a check on between-laboratory bias.

The development in clinical chemistry was parallel and apparently independent. The number of analyses in this field doubles each five years; thus clinical chemists are confronted with the need to cope with this problem. Here the first line of attack was automation, and as is clearly visible when one visits a clinical laboratory, the progress in this field is great. Clinical chemists became aware of the quality problem quite early, even before automation was begun, when confronted with the enormous differences in clinical chemical data between various laboratories. Clinical chemists came to recognize that their laboratories are in many ways similar to a manufacturing plant: reagents and samples are received and subjected to a variety of manipulations using specialized tools and instruments. The final product is the analytical data that are produced.

In 1947 Belk and Sunderman (BE 47) introduced the Shewhart chart as a means to control the product. The conclusions of Belk and Sunderman based on their pioneer survey were later confirmed by the results of similar studies (TO 63, DE 64, ST 65, BE 67). All revealed a wide range of reported values for the analysis of identical specimens among a group of laboratories. In 1950 a formal intralaboratory quality control system, based on estimation of accuracy and precision, was first described by Levey and Jennings (LE 50). Along this line many nationwide, interlaboratory control systems have been established, some under governmental supervision, others under the supervision of boards founded by clinical chemists. However, at present there is no system available that measures quality unambiguously and creates a uniform quality of clinical chemical measurements. In 1979 Limonard suggested the application of control theory to clinical chemistry in order to control quality (LI 79). The vast problems caused by the increasing work load also gave rise to some research on information content, cost, and waiting times (VA 74). This research seems to be independent from the analytical operations research trend described earlier.

One way of defining quality is by stating it to be the (numerical) value of a set of desired properties. In order to control the quality of analytical chemical methods and results, one must define the set of desired properties of the analysis as well as the way to quantify these and the standards to be set.

For practical reasons it is sufficient to consider only those properties that can be quantified and for which standards can be set. Of course it might be possible that more properties are desired, but if they cannot be quantified, it is useless to mention them in the context of this book.

The process of measuring and varying analytical quality can be compared with the control of a technological process. In principle the same procedures can be applied and many of the statistical and control theoretical procedures described elsewhere could be of use. However, an analytical laboratory is far more difficult to control than a chemical process. Some reasons for this can be mentioned:

- Only a few quality parameters are known and fully understood. Measurement of these parameters is often time-consuming and expensive. The number of measurements that can be used for control is often very low, so that "normal" statistics do not apply.
- Most analytical chemical laboratories are relatively labor-intensive, and human effort effects the quality of the work considerably. Quality defects caused by human error are very difficult to control and cannot be coped with completely by control theory.
- The results of analytical chemical measurements affect more than just the institute or factory the laboratory is working for. Products are bought and sold all over the world, and water and air pollution is

transported across boundaries of countries and continents, for example. Therefore the results of analytical laboratories should meet not only laboratory but also national and international standards.

- In analytical laboratories a comparatively large part of the staff consists of highly and very highly skilled and competent workers.

These factors, in combination with the ease and low cost of changing an analytical procedure, are responsible for the astonishing small number of exactly corresponding analytical procedures. Both comparison and control of incomparable procedures on a large scale are difficult.

Nevertheless the "analytical world" has adopted some control instruments that can cope with the situation and use the unique structure of the analytical world. This control mechanism is based mainly on two assumptions:

1. An analyst is able to perform his own control measurements and evaluate them.
2. A laboratory worker can and is willing to improve the quality of his work.

The combination of these seems to be sufficient for control in practical situations. For a more detailed consideration it suffices to distinguish three levels in analytical work that should be controlled. The first level pertains to the unit operation. Some of the operations, for example, sampling, measuring, and data processing, can be evaluated by quantifying desired properties, for example, accuracy or reproducibility. The second level pertains to the procedure itself. Many of the quality parameters described in this book are on the procedure level, for example, sampling frequency, limit of detection, and sensitivity. The third level belongs to the laboratory considered as a unit. When the performance of analytical procedures is evaluated, not only the results of various methods, workers, and instruments are compared, but also the results of other laboratories where comparable procedures are applied. Examples of quality parameters at this level are accuracy, cost, speed, and time lag between question and answer.

This book is subdivided into the three parts of the analytical process: sampling, analysis, and data handling. In addition a chapter on the organization of laboratories is included, for this may have a great impact on analytical quality. As will be seen, the methods used to describe quality are not often applied exclusively to one of these areas. Frequently the techniques have a much wider scope and therefore the concept of measurability, for example, will be found in the section on sampling as well as in the section on analysis.

We feel that the control mechanism that works in many laboratories is reliance on the craftmanship and responsibility of the analyst. Therefore we do not give strict regulations and quality control schemes. Instead we accommodate the analyst with a set of tools, to be applied when and where they are wanted. Since these tools are intended for analytical chemists operating in common practical situations of analytical laboratories, in charge of performing analyses for control and research, we refrain from giving rigorous derivations of the equations presented. Interested readers will find the theoretical background in the original papers cited.

The concept of quality in chemical analysis has not been treated in detail in the literature up to now. Therefore it is not possible to give here a complete treatment of all parameters that influence quality. However, it seems possible to improve quality when the influence of the parameters is known. Another problem is the quality level that is required. Sometimes the level is set by the use that has to be made of the analytical results, and sometimes an optimum exists, depending on other variables. In Section 1.1 a survey is given of quality criteria in chemical analysis.

1.1 QUALITY IN CHEMICAL ANALYSIS

In this section we try to define various quality parameters that are discussed in this book. Because quality in general can be defined as the value of a set of desired properties, it is necessary to describe the parameter that denotes the quality of a certain property of chemical analysis. Furthermore, how quality can be influenced and what features determine the ultimate or optimal quality are discussed.

The quality of a *sample* is given by the similarity between sample and object. Its magnitude can be expressed by the standard deviation in the composition of many samples, or otherwise as the bias in the composition of a sample compared with the "true" value of the composition of the object. Sample quality is influenced by the composition and inhomogeneity of the object, the sample size, and the method of sampling (Sections 2.3–2.5). The upper limit of the quality is set by the precision of the method of analysis that is used to analyze the sample. In addition, the quality of a sample can be influenced adversely by inadequate subsampling, decomposition, and contamination (Section 2.6).

The quality of the *sampling method* is given by the similarity between the reconstruction of the composition of an object and the object itself, as far as it is influenced by the sampling strategy. It depends on the characteristics of the object and the purpose of the reconstruction (Sections 2.2 and 2.3).

For a mere description of the object, the quality can be expressed in the sample quality (Sections 2.5.1 and 2.5.2).

For threshold control of the object, the sampling quality can be expressed in the probability that a threshold crossing will be detected (Section 2.5.3).

For object control the quality can be expressed as controllability or measurability (Section 2.5.4).

Another method for measuring sampling quality is to estimate the effort (expressed in the number of samples to be taken) required to obtain a specified reconstruction error. Here sampling quality is influenced by the inhomogeneity, internal correlation and size of the object, and the number, spacing, and size of the samples. The maximum quality that may be achieved is set by a combination of these factors. The optimal quality is set by the yield of the reconstruction and the cost of sampling. Moreover, the quality of sampling can be influenced adversely by inadequate sample handling, analysis, and data handling (Sections 2.6, 3.2.5, Chapter 4).

The *limit of detection* is a quality parameter pertaining to analysis. It gives the minimum concentration of a component that can be detected. It is influenced by the absolute value of the blank, the standard deviation of the method of analysis, and a safety factor (Section 3.2.1). The lowest possible limit of detection is set by the characteristics of the method.

The *sensitivity* is another quality parameter of the analysis. It gives the change of a signal upon a change of concentration (Section 3.2.2). The sensitivity of a method is a practical quality measure, since it pertains to the ease of detection. However, both the detection of a difference in concentration between two samples and the limit of detection are ultimately governed by the precision of the analytical method and not by the sensitivity. The optimal value of the sensitivity should be such that the standard deviation of the method can be measured.

Selectivity and specificity are quality parameters of a method of analysis. They can be expressed as the ratio of sensitivities of the method for various components to be measured in the sample. The optimal value of these characteristics is set by the number of measurements required to estimate the composition of the sample (Sections 3.2.8, 3.2.9).

Safety is an important but ill-defined quality parameter of a method of analysis. Here hardly anything is known about quality criteria. Perhaps the cost of measures to be taken to satisfy safety regulations can be used to measure safety as a quality parameter (Section 3.2.3). The optimum is determined by the balancing of adverse consequences and precaution costs.

Cost is an important quality parameter of an analytical method. However, there are hardly any publications on the subject. Cost can be expressed in money or in other measures, such as manpower. It is influenced by analysis time, standard of labor, depreciation, and cost of instruments, energy, and reagents (Section 3.2.4). The optimal cost of an analytical method results from a comparison of various methods of analysis with respect to the yield of the object and cost of analysis (Section 2.5.4).

Precision is a quality parameter that, important as it may be, is not well defined. *Analytical Chemistry* (AN 75) proposes the following: "Precision refers to the reproducibility of measurement within a set, that is, to the scatter or dispersion of a set about its central value." The term "set" itself is defined as referring to a number of independent replicate measurements of the same property. The International Organization for Standardization (ISO) (IN 66) applies two descriptions of precision: (1) the *reproducibility*, the closeness of agreement between individual results obtained with the same method but under different conditions, and (2) the *repeatability*, the closeness of agreement between successive results obtained with the same method and under the same conditions.

Both descriptions of precision use the same measure, the standard deviation of the measurements. Apart from the influence of external conditions, the precision is influenced by the number of replicate measurements (Section 3.2.6). The ultimate precision of analysis is determined by the precision of the method and the quality of the sample and sampling (Sections 2.5.2–2.5.4). The optimal precision is determined by cost, sample, sampling, and aim of the analysis (Sections 2.5.4 and 3.2.4).

Accuracy is an even less defined quality criterion. *Analytical Chemistry* (AN 75) states: "Accuracy normally refers to the difference (error or bias) between the mean of the set of results and the value which is accepted as the true or correct value for the quantity measured." Because most analytical methods claim to give the true value, accuracy as such is seldom used as a quality criterion (Section 3.2.7). Often the difference between the mean of the set of results and the mean of a much larger set is taken for accuracy (Sections 3.4.3 and 4.6).

Information can be considered as a quality parameter for quantitative and qualitative analysis and for data retrieval. Information is defined as the difference in uncertainty before and after an experiment. It is expressed as the binary logarithm of this uncertainty and measured in bits. The information content of a qualitative analytical method is governed by the selectivity and specificity of the method and the *a priori* knowledge of the occurrence of the sought component. The information content of a quantitative analytical method is a function of the standard deviation of the method and the *a priori* knowledge of the range of compositions that can be expected. The information content of a retrieval method is governed by the selectivity of the method and the chance of occurrence of the sought item. Therefore information cannot be a quality criterion for a method as such. It can be applied only in a specified situation. Correlation of items makes the *a priori* information partly predictable and thus diminishes the information content of the experiment (Section 4.2).

Curve fitting is not a quality parameter, but it may enhance the quality of measured data considerably for two reasons:

1. Interpolating the gap between data allows easier interpretation of many phenomena, although the information does not increase.

2. A most important use of curve fitting is in the unraveling of mixed phenomena, for example, spectra (Section 4.4).

Pattern recognition may be a tool in establishing new quality criteria. The first application may be in discovering patterns in the data produced by analytical chemists. This enhances the value of these results. A measure for this effect is not known, but a "patterned" result seems to increase the quality of the analytical results. A second application may be in revealing the weight of various measured parameters in recognizing a pattern. The quality could be measured by the ratio of required and obtained analytical information. A third application may be found in the selection of an analytical method for a certain problem. The quality parameter here could be the similarity of the method chosen by intuition or experience and the method selected by pattern recognition. At the moment it can only be said that pattern recognition may influence the quality of an analytical method (Section 4.7).

Optimization is not a quality parameter. However, it may be applied in optimizing analytical methods and thus improving quality. Perhaps the difference between a method as such and a method optimized with respect to a number of quality parameters could be a quality criterion (Section 4.8).

1.2 QUALITY CONTROL

Quality of chemical analysis can be improved or kept on a steady level by controlling the parameters that influence quality. The most direct way is to control the parameters that are given by the equations that describe quality quantitatively. Although many quality parameters cannot be described quantitatively, nevertheless quality may be influenced. A number of techniques that influences or may influence quality are described in the book.

Sampling as a means to influence quality was described in Section 1.1.

Handling of samples can be of great importance for some quality aspects of a sample. Homogenizing influences the quality of the subsample (Section 2.6.1) and so does sample reduction (Section 2.6.2). Preservation and labeling influence both the identity and integrity of the sample, an immeasurable quality aspect (Section 2.6.3).

The *selection* of an analytical method influences many quality aspects. The selection may be based on one or more quality criteria and therefore these aspects should be weighed *a priori* (Section 3.3). Techniques for selection are, for example, the ruggedness test (Section 3.4.1), the ranking test (Section 3.4.4), pattern recognition (Section 4.7), and optimization (Section 4.8).

Precision control and accuracy control can be performed in several ways. Control charts are means to trace quality in time and indicate when intervention is required (Section 3.4.2). Round robin tests may reveal shortcomings in precision and accuracy and, though complicated psychological and organizational influences play a role, quality can improve after a test (Section 3.4.3). Sequential analysis (Section 3.5) as well as analysis of variance (Section 4.6) may be used to control precision and accuracy. Neither method influences the quality parameters but both may be of help in deciding which method is preferred. A number of methods can be used to improve precision by removing or diminishing irrelevant data (noise). Some techniques treated are curve fitting and smoothing (Section 4.4), Kalman filtering, and deconvoluting filtering (Section 4.3). To improve data handling and quality control by statistical means, often a mathematical transformation of data may be helpful (Section 4.1.2).

Planning may improve some quality aspects, for example, speed and cost; by planning over a longer time span other quality aspects may be influenced by research. Speed may be improved by the optimal use of available manpower and instruments. Dynamic modeling and queueing theory provide tools for simulation on laboratory models (Section 5.1.1). Gantt charts and network diagrams provide the analyst with simple means to avoid dead times (Section 5.1.3).

The availability of methods that have a better quality than the existing methods may be forecasted or promoted by research. Examples of forecasting methods are trend extrapolation and dynamic and intuitive modeling (Section 5.1.1). Promotion of research by increasing creative thinking is possible by, for example, needs analysis, lateral thinking, and Synectics®

(Section 5.1.1). Promotion of research through research planning can be supported by network diagrams and flow charts, for instance (Section 5.1.3).

The *organization* of a laboratory influences the quality of chemical analysis in many ways, mostly complicated and obscure. The most evident influence is found on speed and cost, but accuracy and precision and the entire subject of quality control may be advantageously or adversely influenced by the organization of analytical work. Therefore, qualitative as it may be, a section on organizational aspects is included (Section 5.2).

REFERENCES

AL 69 P. L. Alger: *Mathematics for Science and Engineering*, McGraw-Hill, New York, 1969.

AN 75 "Analytical Chemistry," *Anal. Chem.* **47**, 2527 (1975).

AR 73 Arbeitskreis: "Automation in der Analyse," *Talanta*, **20**, 811 (1973).

BA 28 B. Baule, A. A. Benedetti-Pichler: *Z. Anal. Chem.* **74**, 442 (1928).

BE 47 W. P. Belk, F. W. Sunderman: *Am. J. Clin. Pathol.* **17**, 854 (1947).

BE 67 R. E. Berry: *Am. J. Clin. Pathol.*, **47**, 337 (1967).

BE 75 R. J. Bethea, B. S. Duran, T. L. Boullion: *Statistical Methods for Engineers and Scientists*, Marcel Dekker, New York, 1975.

DA 72 O. L. Davies, P. L. Goldsmith: *Statistical Methods in Research and Production*, Oliver and Boyd, Edinburgh, 1972.

DE 64 F. B. Desmond: *N. Z. Med. J.* **63**, 716 (1964).

DI 69 W. J. Dixon, F. J. Massey: *Introduction to Statistical Analysis*, 3rd ed., McGraw-Hill, New York, 1969.

DU 74 O. J. Dunn, V. A. Clark: *Applied Statistics: Analysis of Variance and Regression*, Wiley, New York, 1974.

FI 55 D. J. Finney: *Experimental Design and its Statistical Basis*, Cambridge University Press, 1955.

FI 70 A. F. Findeis, M. K. Wilson, W. W. Meinke: *Anal. Chem.* **42**(7), 26A (1970).

GO 08 W. S. Gosset: *Biometrika*, **6**, 1 (1908).

GI 63 P. M. E. M. van der Grinten: *Control Eng.* **10**(12), 51 (1963).

GR 63 P. M. E. M. van der Grinten: *Control Eng.* **10**(10), 87 (1963).

GR 65 P. M. E. M. van der Grinten: *J. Instrum. Soc. Am.* **12**(1), 48 (1965).

GR 66 P. M. E. M. van der Grinten: *J. Instrum. Soc. Am.* **13**(2), 58 (1966).

IN 66 "International Organization for Standardization," Technical Committee 69, ISO/TC 69, 1966.

KA 79 G. Kateman, A. Dijkstra: *Z. Anal. Chem.* **247**, 249 (1979).

KO 72   B. R. Kowalski, C. F. Bender: *Anal. Chem.* **44**, 1406 (1972).

KO 75   B. R. Kowalski: *J. Chem. Inf. Comput. Sci.* **15**, 201 (1975).

LE 50   S. Levey, E. R. Jennings: *Am. J. Clin. Pathol.* **20**, 1059 (1950).

LE 71   F. A. Leemans: *Anal. Chem.* **43**(11), 36A (1971).

LI 79   C. B. G. Limonard: *Clin. Chim. Acta* **94**, 137 (1979).

MA 75   D. L. Massart, L. Kaufman: *Anal. Chem.* **47**(14), 1344A (1975).

MA 77   D. L. Massart, H. de Clerq, R. Smits: "Reviews on analytical Chemistry," lecture presented at *Euroanalysis Conference II*, Masson, Paris, 1977, p. 119.

NA 63   M. G. Natrella: *Experimental Statistics*, National Bureau of Standards Handbook 91, U.S. Govt. Printing Office, Washington D.C., 1963.

SH 31   W. A. Shewhart: *Economic Control of the Quality of Manufactured Product*, Van Nostrand, New York, 1931.

ST 65   J. V. Straumfjord, B. E. Copeland: *Am. J. Clin. Pathol.* **44**, 252 (1965).

TO 63   D. B. Tonks: *Clin. Chem.* **9**, 217 (1963).

VA 74   I. R. Vaananen, S. Kivirikko, J. Koskenniemi, J. Koskimies, A. Relander: *Methods Inf. Med.* **13**(3), 158 (1974).

VA 77   B. G. M. Vandeginste: *Anal. Lett.* **10**, 661 (1977).

WI 79   A. L. Wilson, *Analyst* **104**, 273 (1979).

YO 51   W. J. Youden: *Statistical Methods for Chemists*, Chapman and Hall, London, 1951.

SAMPLING

Sampling is often called the basis of analysis: "The analytical result is never better than the sample it is based on." In our opinion, every part of the analytical procedure is important, because the final result is influenced by mistakes or added noise in every part of the procedure. Of course this statement simply emphasizes the need to control the sampling as well as every other part of the analytical work.

The purpose of sampling is to provide for a specific aim of the patron a part of the object that is representative of it and suitable for analysis. The quality of the sampling depends on many parameters, governed by the properties of the object and of the analytical procedure. The properties of the object can be described in terms of physical structure, size, and inhomogeneity in space and time.

The properties of the analytical procedure can be described in terms of minimum and maximum sample size, physical structure of the sample, homogeneity of the sample in space, and stability in time, for example.

The requirements set by the object and by the analyst must be fulfilled, so the sampler has to optimize the sampling parameters such as size, time or distance spacing, number of subsamples, cost, mixing, and dividing.

In this chapter we deal with these parameters, their influence on the quality of the sample, and the procedures to be used.

2.1 DEFINITIONS

2.1.1 Quality

A definition of quality is "the (numerical) value of a set of desired properties."

The estimation of the value of the properties is the aim of the analytical procedure. A feature of the object to be known is a list of the desired properties, for these determine the way of sampling.

The object can have many properties, for example, composition, particle size, color, and taste, but only a selection of these properties is required to describe the object quality. In assessing the quality of coal it makes a difference whether the desired property is "ash content" or "particle size distribution," though in both cases the analyst as well as the sampler must consider the proper sampling of the different fractions of different particle size.

One sometimes comes across the term "total analysis." Most often this refers to a known list of desired properties, but sometimes "total analysis" simply means as many properties as can be estimated.

A stain of paint found on a shirt has quality properties that are different from the tin of paint it comes from. Though it is possible to define object quality in this way, its implication is that the analyst or sampler must define the object quality afterward: analyzing the stain with respect to pigment composition, binder, color, or degree of polymerization, for example, is possible by analyzing a proper sample of the stain. The object qualities "weight" or "solvent content" can be established only by sampling the stain in a different way.

2.1.2 The Object

Before sampling and analysis some aims have to be set: what is the object that has to be described and what quality* parameters are wanted? We define an object here as "the entity to be described."

An object can be a fertilizer granule as well as a bag or truckload of fertilizer, or even the quantity produced last week. An object can be a river in its full length during one year, or the water flowing past a given point during a given time, or the water at one point at a given moment.

As a rule the object is sufficiently described by the four coordinates of space and time, though in practice other labels may be more useful.

Other variables that are required in order to describe the objects are the state variables temperature and pressure, for often the sample cannot be kept in the same condition.

2.1.3 The Sample

Before describing "sample qualities," we define sampling and samples. A useful working definition of a sample is "a representative part of the object to be analyzed." Because representativeness is needed only for the object quality parameters, it can as well be stated, "A sample is part of the object selected in such a way that it possesses the desired properties of the object." As shown later the sample can resemble the object only to

*To avoid confusion, in this chapter the word "quality" is not used on its own, as in normal use, but always coupled to an adjective, as in sampling quality or object quality.

a certain degree, but because this degree is set by the method of analysis, it can be omitted in the definition of "sample."

A sample the size of a truckload can be a good sample—that is, representative of the object—but it cannot be used in the laboratory. A very small sample can be as good a sample as a larger one, but perhaps it is impossible to handle. To make the definition useful in practice the following criterion must be added: "The sample must have such dimensions that it can be analyzed."

Another criterion that has to be added is, "The sample must keep the properties the object had at the time of sampling, or change its properties in the same way as the object." Thus deterioration of the sample through exposure to the air, the sample container, or microorganisms, for example, should be avoided.

The sample procedure or "sampling" is a succession of steps performed on the object which ensures that a sample possesses the specified sample quality.

2.2 PURPOSE OF SAMPLING

Because sampling depends on the purpose of sampling, the various purposes must be defined.

2.2.1 Description

One purpose of sampling may be the collection of a part of the object that is sufficient for the description of the gross composition of the object, for example, sampling lots of a manufactured product, lots of raw material, or the mean state of a process. Here it is desirable to collect a sample that has the minimum size set by the condition of representativeness or demanded by handling.

Another purpose of sampling can be the description of the object in detail, for example, the composition of a crystallized rock as a function of space or the composition of the various particles in a mixed pigment. Here it is necessary to know the size and the number of samples or the distance between the samples (sampling frequency).

2.2.2 Threshold Control

Here the purpose of sampling is to collect a quantity of sample sufficient for guarding the object. This situation differs from that of control. In guarding it is often sufficient to know that a threshold value has been reached, or will be reached with a certain probability. If action is required, it can be approximate. When the percentage of active component of a catalyst is monitored, for example, the action is the replacement of the catalyst when the threshold level has been reached, or even before. The guarding of the effluent of a factory may have the same characteristics: when a certain level is reached or will be reached within due time, the effluent stream is stopped.

Another characteristic of threshold control is that in most cases only one level is monitored, sometimes the level pertaining to the high limit threshold and sometimes the low one. This is contrary to the situation met with control, when deviation to either side should be eliminated. In threshold control the quality parameter can be the optimal number of samples required for guarding the process, or the minimum number of samples required to predict when the process will reach the threshold value. The time span between the warning that the process will reach the threshold level and the actual crossing of that level also can be a quality parameter.

2.2.3 Control

In industrial practice most of the analytical effort is dedicated to process control. The purpose of control is to keep a process quality constant (e.g., a manufacturing process), as close to a desired value as is technically possible and economically desirable. Part of the deviation from the set point is caused by random fluctuations of process conditions. These fluctuations cause the composition of the sampled product to vary at random between certain limits. In order to control the fluctuating process, samples must be taken with such frequency, and analyzed with such reproducibility and speed, that control actions can have the optimal result.

A second purpose of sampling can be the detection of nonrandom deviations: drift or cyclic variations. Special techniques are required to discover such anomalies as soon as possible. These also set the conditions for sampling frequency and sample size.

2.3 TYPES OF OBJECTS AND SAMPLES

Objects can be divided into three groups according to the scheme of Table 2.1. The discrete or continuous changes in object quality can manifest themselves in time as well as in space.

The composition of the output of a factory changing with time can be seen as a time-related property by sampling the output. It can be seen as a

TABLE 2.1.   Types of Objects

Object		
Homogeneous No change of quality throughout the object	Heterogeneous	
	Discrete Change of quality throughout the object	Continuous Change of quality throughout the object
Homogeneous	Discrete Changes	Continuous Changes
Well-mixed liquids Well-mixed gases Pure metals	Ore pellets tablets Crystallized rocks Suspensions	Fluids or gases with gradients Mixture of reacting components Granulated materials with granules much smaller than sample size

space-related property by sampling the conveyor belt or the warehouse. In many cases such changes are common. Therefore both kinds of changes are treated as equal. Examples of the three types of objects are given in Table 2.1.

These types can be distinguished only at the macroscopic level. When objects the size of molecules are considered all objects are heterogeneous with discrete quality changes.

2.3.1   Homogeneous Objects

The sampling of homogeneous objects poses fewer problems than the sampling of other types. Because the definition of homogeneous is "the same composition at every point and every moment," it is clear that any grab taken without bias is a sample. However, true homogeneous objects are rare, and homogeneity is proved only after the hypothesis "this object is homogeneous" is verified by sampling in such a way that the hypothesis is not used as an axiom.

2.3.2   Heterogeneous Objects

Both types, the objects with discrete changes and those with continuous changes, are distinguishable only at a certain level. Because analytical methods have a finite accuracy, quality cannot be discriminated better than ts.* Therefore it makes no difference whether the object is considered

---

*t = Student's-t for the appropriate number of measurements;
s = standard deviation of the method.

to be of the type "discrete changing in quality" when quality changes are less than or equal to *ts*.

Three types of heterogeneous objects are possible:

1. *Random or Pseudorandom Distribution of the Quality Parameters.* This type of object is rare. As a rule these objects are randomly composed of other objects. Examples are the population in an area and a lot of discrete objects collected from sources randomly chosen, such as tins of food of different brands.

2. *Correlated Distribution of Random Variable Quality Parameters.* This type of object is the most common. Examples are the different parts of a geological area, the output of a chemical process, and the composition of a large basin of water or a flowing river. The degree of correlation can be described by techniques such as correlation analysis and auto- or cross-correlation of time or distance series.

3. *Cyclic Correlated Distribution of Object Quality Variables.* Examples are the annual changing composition of living organisms or rivers, for instance, and the time-dependent variations of the output of factories (shift-to-shift variations, day-to-day temperature or light-induced cycles). This type of correlation can be described by Fourier analysis or autocorrelation techniques.

Many frequencies are often found, the shift frequency, day frequency, and a control-induced frequency occurring mostly in industrial processes. In natural objects frequencies correlated with day and season are normal. The three types of heterogeneous objects are seldom found in their pure state; mixed forms with one or two dominating types are most common.

2.3.3 Types of Samples

Often samples are made up from increments or grabs. As long as these increments are real samples, used to build up the sample to increase its size, they are called samples used to build up a gross sample. However, when these increments are parts of the object but not representative of the object, they are called, not samples, but sample increments or grabs.

The sample or gross sample can be subdivided into smaller samples. When the properties of these samples are equal to those of the gross sample, or have the required sample quality, they are called subsamples. Note that a subsample by definition is a real sample! An analysis sample is a subsample or sub-subsample used for analysis. It is the only member of the sample family that must fulfill the added criterion of required dimension.

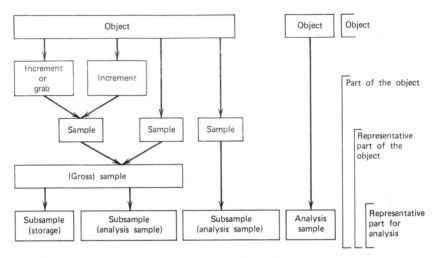

Figure 2.1. Sample nomenclature.

Other samples, as a rule being subsamples with the same sample quality, can be used for different purposes, such as duplicate analyses, referee analysis, and storage. This leads to the scheme of Figure 2.1.

Often a sample is not a separate part of the object but the whole object, for example, very small objects or objects to be analyzed by nondestructive methods. Sometimes a part of the object is declared to be a sample, for example, when analyzing a steel pipe with emission spectroscopy *in situ*. The same definitions hold, although not all kinds of samples can exist for these cases.

2.4 SAMPLE QUALITY

Sample quality can be defined as "the (numerical) value of the set of desired properties of the sample." Sample quality is to be distinguished from object quality. Object quality is the set of properties that must be estimated in the analytical procedure. Sampling quality is the set of properties of the sample that leads to a good sample.

The primary sample quality parameter is representativeness. Secondary parameters are size, stability, and cost. For changing objects, additional parameters are discriminating power and/or speed.

2.4.1　Representativeness

The sampling procedure must fulfill the condition that analysis (with the required accuracy) of the sample shows no difference in object quality compared to the object itself. This means that two samples of the same object cannot be distinguished. In practice this condition is often not met. The result is that the accuracy of the total analysis (including sampling and analysis) is diminished to such an extent that the aforementioned condition is fulfilled. Two cases can be distinguished: the sampling is prone to random variations, or the sampling causes systematic deviations. In Section 2.5 the connection between reproducibility of the method of analysis and method of sampling is derived.

2.4.2　Systematic Deviations in Sampling

Some causes of systematic deviations in sampling are the following:

1. The number of increments is too small, so that the sample shows a bias. In fact this is a random deviation.
2. The sampling procedure is preferential to one or more object quality parameters. Continuous sampling of air or water can be biased for particulates because of the different inertia of the particulates compared to the bulk. Sampling of granulated product at a conveyor switch point can miss the fines, which escape as dust. Electrostatically charged particles tend to clog or escape, as can living organisms.
3. The sampling procedure causes alterations in the object. Obvious causes are crushing of granules during sampling, evaporation, and segregation, for example. More complicated are sampling procedures that trigger a reaction in the object, such as catalytic decomposition by the sampling tool, oxidation by oxygen introduced during sampling, fear-induced alterations of zoological samples, and changing reactions of living cells, for instance, the potassium "bleeding" of cells after venopuncture.
4. The sample changes after the sampling procedure and before the analysis. Examples are the oxidation, dehydration, or biological degradation of samples, coagulation, and condensation. A distinct cause is the interaction with the sample container by corrosion, adsorbtion, or catalytic action.
5. The sample is altered intentionally. The results of analysis can have strong economic or psychological repercussions. Because often during the process of sampling humans can or must interfere, and because the

sampling is nearer to the involved party than analysis is, sampling is liable to intentional alterations. Bribery, the fear of being accused of malproduction, and the more or less "esthetic" desire to show the object at its best are the more common causes.

2.5 SAMPLING PARAMETERS

Since the sampling procedure affects the sample, the so-called sampling parameters that define this procedure must be considered. The object or process is represented by the distance $0-t$ in Figure 2.2. It is an ergodic time series, which means a process than can be represented mathematically by a time series, consecutive numbers representing, say, composition as a function of time, of such a length that the properties of the time series—mean, standard deviation, and autocorrelation—are independent of the length. In this series a lot can be represented by the distance P, a nonergodic time series.

A sample is the sum of n sample increments or grabs, taken during a time span G with a frequency $1/A$. In the rest of this chapter the influence of the parameters, G, A, and n on the estimation of the composition of the lot P or the process is considered for various types of lots.

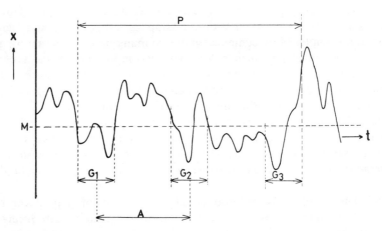

Figure 2.2. Lot size P, grab size G, and distance between samples A as a function of time or distance.

2.5.1 Samples for Gross Description

Homogeneous Objects

If the object to be sampled is homogeneous, for example, a well-mixed tank or lake, it is obvious that here every sample, whatever its size, is a true copy of the object. Theoretically one sample suffices. Often in practice more samples are needed, for instance, to fulfill the requirement that the sample must have a minimum size. When the sample size is too large to be obtained in one sample, such as in precious objects of art, more samples are needed. The size of the gross sample, set by the requirements of the analysis, can be estimated as follows: Assume the sample weight W and the fraction of the component to be estimated P ($1\% \rightarrow P = 0.01$, 1 ppm $\rightarrow P = 10^{-6}$). Now the total amount of the component in the sample is $T = PW$. A yardstick for sample size estimation is provided by the nomogram of Figure 2.3, which is based on the use of the logarithmic operator $p = -\log$. This implies that $pT = pP + pW$ (AR 72).

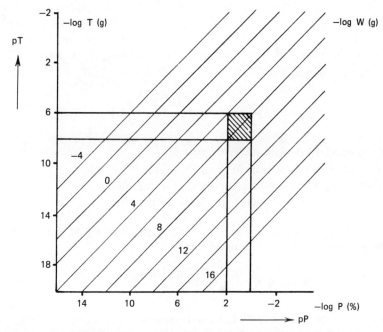

Figure 2.3. Nomogram for the interdependence of sample size W, composition P, and total amount of component T (AR 72).

Heterogeneous Objects

A situation that is often encountered is the sampling of an object that consists of two or more particulate parts. Since the first treatment of the subject by Baule and Benedetti-Pichler (BA 28), many authors have been engaged in the subject (AS 56, DU 62, SK 70, WI 64, GY 66, VI 69, DU 71, VI 71, IN 73, HA 74, HA 76, IN 76, KL 77, GY 79). Most of these papers are refinements of the original theory. Here only the general case is treated.

Suppose an object consists of two granular components, A and B; the particles are spherical granules of identical size and density. Now it is possible to predict the composition of a grab or increment. The composition of an increment depends on the size of the increment, for two different reasons. The particulate composition is responsible for a discontinuous change in composition. If the composition is 50% A and 50% B, say, an increment of one granule will have the composition either 100% A or 100% B. An increment of two granules can have the composition 0, 50, or 100% A; a three-particle grab can be 0, 33, 67, or 100% A.

The cause of difference between the grab and the object is the inevitable statistical error in taking a grab from the object. The composition of the grab, the collection of increments that must ultimately constitute a sample, is given by m and s, m being the mean composition of the object and s the standard deviation of this mean. From statistics it is known that $s = \sqrt{P_A P_B n}$ particles or

$$s_g = \frac{1}{n} \sqrt{P_A P_B n} \times 100 = 100 \left(\frac{P_A P_B}{n} \right)^{1/2} \% \qquad (2.1)$$

with s_g = standard deviation (%)
  P_A = fraction of component A in the object
  $P_B = 1 - P_A$ = fraction of component B in the object
  n = number of increments (particles) in the grab

Assume the same composition as before, 50% A, 50% B, $P_A = 0.50$, and $P_B = 0.50$. A grab with $n = 10$ will have a composition

$$m = 50\% \text{ A}, \qquad s_g = 100 \left(\frac{0.5 \times 0.5}{10} \right)^{1/2} = 16$$

Two out of three grabs will have a composition between $50 \pm 16\%$, or between three and seven particles A. The probability that a grab is found outside this region is one to three. In fact the probability of finding a grab with the composition 10 particles A or B is not negligible: 1% of the grabs will have a composition of 0 or 10 particles of one kind.

From eq. 2.1 it follows that the chance of a deviating composition diminishes with increasing n. If the grab is considered to be representative of the object, when there is no means to distinguish this grab from another one taken from the object, a method of analysis is required that has an ultimate accuracy larger than the standard deviation of the grab.

When the accuracy of the analysis is $s_a = s_A / \sqrt{N}$, where
s_A = standard deviation of the method of analysis and
N = number of repeated analyses, no distinction can be found between two grabs provided $s_g < s_a$. In other words, it is known *a priori* that a grab will not be equal to the object. This is not important, however, if the difference cannot be seen.

The condition set for a grab to be a sample is $s_g \leqslant s_a$ or

$$100\left(\frac{P_A P_B}{n}\right)^{1/2} < s_a \qquad (2.2)$$

which is equivalent to

$$n > \frac{P_A P_B \times 10^4}{s_a^2} \qquad (2.3)$$

In the example of 50% A and 50% B a condition can be an accuracy of 1% for the determination of A in a mixture of A and B. Assuming that every analysis is executed in duplicate, $s_a = 1/\sqrt{2} = 0.7\%$, then

$$n > \frac{0.5 \times 0.5 \times 10^4}{(0.7)^2} = 5100$$

Application of eq. 2.3 shows that a grab consisting of at least 5100 increments is required before it can be considered as a sample, a representative part of the object. If the method of analysis is less accurate, say, 10%, n diminishes rapidly, being

$$n > \frac{0.5 \times 0.5 \times 10^4}{10^2} = 25$$

Even without any refinement of the theory, such as the introduction of different densities for A and B and nonuniform particle size distribution, some important remarks can be made on the act of sampling. Because the size of the sample is expressed in number of particles, the weight of the sample depends on the weight per particle.

For a coal/bedrock mixture with an average unit weight of 100 g per particle, 5100 increments have a total weight of 510 kg, as for a mixture of two pigments with a particle weight of $\sim 10^{-11}$ g a sample of 1 mg is adequate, if only sample composition is considered. The coal sample of 500 kg, when crushed to a mean of 1 g per particle, can be sampled by a subsample of 5100 g or just 5 kg.

The preceding theory is based on a homogeneous distribution of the components in the object and nonbiased sampling, or a random sampling when the object is not homogeneous. In practice it is difficult to homogenize the object when it is more than about 100 kg. In obtaining a subsample from a gross sample, as a rule it is possible to homogenize the sample. How to obtain a random sample is treated below.

The foregoing theory does not apply solely to granular objects. Every object, solid, particulate, liquid, or gaseous, can be thought of as consisting of units of equal weight but of different composition.

In order to deal with a different composition, instead of two different entities A and B, the theory must be extended. Instead of speaking in terms of percent particles A, the composition can be expressed in percentage of the a component of both A and B. If C_A is the percentage of a component in A and C_B in B, and no other components are present, the mean composition of a grab is

$$C = \frac{nP_A C_A + n(1 - P_A)C_B}{n} = P_A C_A + (1 - P_A)C_B \qquad (2.4)$$

If the grab consists solely of component A, the composition is C_A, and if soley of B, C_B. The mean standard deviation in the composition is

$$s_m = 0.01|C_A - C_B|s_g = |C_A - C_B|\sqrt{\frac{P_A P_B}{n}} \qquad (2.5)$$

Applying the conditions $s_m \leqslant s_a$ for a sample yields

$$n > \frac{P_A P_B (C_A - C_B)^2}{s_a^2} \qquad (2.6)$$

If in the example A contains 10% Fe and B 5%, and if it is assumed that $P_A = P_B = 0.5$, the duplicate analyses having an accuracy of 1%, then

$$n > \frac{0.5 \times 0.5 \times (10 - 5)^2}{1^2}$$

so $n > 6.25$ or at least seven particles. When, however, the accuracy of the analysis s_a is 0.1%, n increases to 625.

Note that n is at a maximum for $P_A = 0.5$, other values of P_A causing less variance in the composition, so smaller samples can be used. However, lower values of C or m, the mean composition of sample or object, as a rule require the introduction of better methods of analysis in order to keep the relative standard deviation at a reasonable level. Figure 2.4 shows the relation between the number of units required to keep sampling errors within predetermined limits (HA 74). When the relative standard deviation $s_R = s_a/c$ has been used instead of s_a,

$$n = \frac{P_A(1 - P_A)(C_A - C_B)^2}{s_R^2} \tag{2.7}$$

The percent relative difference in composition $100(C_A - C_B)/C_A$ is used to

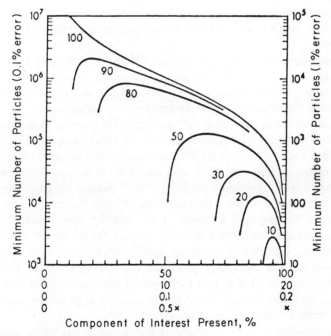

Figure 2.4. Relation between the minimum number of units in a sample required for sampling errors (relative standard deviations in percentage) of 0.1 and 1% (y-axis) and the overall composition of a sample (x-axis), for mixtures having two types of particles with a relative difference in composition ranging from 100 to 10% (HA 74). Reprinted with permission from *Anal. Chem.* **46**, 314 (1974). Copyright by the American Chemical Society.

label the curves so that any range of absolute values on the horizontal axis can be chosen.

Figure 2.4 indicates that sampling error or sampling size decreases sharply as the relative difference in percentage of the substance of interest in the two kinds of particles becomes smaller.

The density of the particles can be incorporated by using

$$n > \frac{P_A P_B (C_A - C_B)^2}{s_a^2} \left(\frac{d_1 d_2}{d^2} \right) \tag{2.8}$$

with $d_1 = d_2$ = density of the two types of units making up the sample and d = overall sample density. For minor and trace elements, assuming the component sought is concentrated in one kind of particle $P_B \to 1$, $C_B = 0$.

Internally Correlated Objects

Objects can be internally correlated in time or space; for example, the composition of a fluid emerging from a tank does not show random fluctuations, but is correlated to the composition in earlier or later sampled product.

As mentioned in Section 2.3, factors causing the internal correlation of objects include (1) diffusion or mixing within the object, for example, in mixing tanks, buffer hoppers, and rivers and (2) varying properties of the producer of the object, reactors or emitters, for instance.

In both situations samples are mutually dependent. When an object shows a large internal correlation, two adjacent samples do not differ much from each other. The difference between two samples increases with greater distance, however. One may ask how to estimate the number of grabs that have to be taken to get a "real" sample or gross sample. As stated for the case of the heterogeneous objects, the number of grabs n depends on the required variance of the sample, this variance being so small that the difference between two samples cannot be detected.

The sample size is influenced by many factors, including the lot size. Unfortunately, the equations that describe the variance of the sample cannot be simply altered to give the required number of grabs, but must be estimated by iterative methods. Another method is to derive them from a graphic representation of the equations.

When the object to be analyzed is a process, a stream of material of infinite length with properties varying in time, the sampling parameters can be derived from the process parameters. When the object is a finite part of a process, however, usually called a lot, the description of the real composition of the lot depends not only on the parameters of the process the lot is derived from, but also on the length of the lot. The sample now must represent the lot, not the process (Figure 2.2).

Lots derived from Gaussian, stationary, stochastic processes of the first order allow a theoretical approach. In practice most lots seem to fulfill the above requirements with sufficient accuracy to justify the following equations (MU 78, KA 78, ME 67). An estimate of the mean m of a lot with size P can be obtained by taking n samples of size G, equally spaced with a distance A. In this case $P = nA$ and the size of the gross sample $S = nG$. Here it is assumed that it is not permissible to have overlapping samples and that $A \geqslant G$.

The relevant parameter is the variance in the composition of the gross sample compared to the variance of the lot. This variance σ_*^2 is thought of as composed of the variance in the composition of the gross sample itself, σ_m^2, the variance in the composition of the whole lot σ_μ^2, and the covariance between m and μ:

$$\sigma_*^2 = \sigma_m^2 + \sigma_\mu^2 - 2\sigma_{m\mu} \tag{2.9}$$

As is shown in MU 78,

$$\sigma_m^2 = \frac{2\sigma_x^2 T_x^2}{nG^2} \left\{ \frac{G}{T_x} - 1 + \exp\left(-\frac{G}{T_x}\right) + \left[\exp\left(-\frac{G}{T_x}\right) + \exp\left(\frac{G}{T_x}\right) - 2 \right] \right.$$

$$\left. \times \left(\frac{\exp(-A/T_x)}{1 - \exp(-A/T_x)} - \frac{\exp(-A/T_x)[1 - \exp(-P/T_x)]}{n[1 - \exp(-A/T_x)]^2} \right) \right\} \tag{2.10}$$

$$\sigma_\mu^2 = \frac{2\sigma_x^2 T_x^2}{P^2} \left[\frac{P}{T_x} - 1 + \exp\left(-\frac{P}{T_x}\right) \right] \tag{2.11}$$

$$\sigma_{m\mu}^2 = \frac{\sigma_x^2 T_x^2}{nPG} \left\{ \frac{2nG}{T_x} + \left[1 - \exp\left(-\frac{P}{T_x}\right) \right] \right.$$

$$\left. \times \left(\frac{\exp(-G/T_x) - 1}{1 - \exp(-A/T_x)} + \frac{\exp(G/T_x) - 1}{1 - \exp(A/T_x)} \right) \right\} \tag{2.12}$$

in which σ_x^2 = variance of process and T_x = correlation factor of process. As can be seen in eqs. 2.9–2.12 σ_*^2 depends on a number of factors.

The properties of the process from which the lot stems are described by σ_x, the standard deviation of the process, and T_x, the correlation factor of the process.

The only relevant property of the lot here is its size P, expressed in the same units as T_x. These units may be time units, such as when T_x is measured in hours. In this case T_x is usually called the time constant of the process. The lot size is expressed in hours as well. When describing a river, the lot size can be the mass of water that flows by in 1 day or year.

The correlation factor T_x can be called a space constant when the unit used is length, however. Here the lot size is expressed in length units. This is the case, for example, when a lot of manufactured products contained in a conveyor belt or stored in a pile of material in a warehouse is considered.

The correlation factor also can be expressed in dimensionless units, as when bags of products are produced. In this case the lot size is also expressed in terms of items (number of bags, drums, tablets). The properties of the sample are the grab size G, expressed in the same units as T_x and P, the distance between the middle of adjacent grabs A (also in the units of T_x, P, and G), and the number of grabs n that form the sample. As is obvious P, G, and A can be expressed in dimensionless units when they are divided by T_x.

The units $p = P/T_x$, $g = G/T_x$, and $a = A/T_x$, together with $F = G/P$, the sample size as fraction of the lot size, are used in Figure 2.5, which represents graphically the rather awkward equations 2.9–2.12.

If the sample size is chosen such that the subsequent samples are taken without interruption, it can be shown that $\sigma_*^2 = 0$. This result is not surprising because now the whole lot has been sampled: $A = P/n$, and the sample size $nG = P$. For an uncorrelated lot, with $T_x = 0$, it can be shown that σ_m, σ_μ, and $\sigma_{m\mu}$ and consequently σ_* are all 0. If, however, the sample size $G = 0$, $\sigma_m^2 = \sigma_x^2/n$ and $\sigma_*^2 = \sigma_x^2/n$.

In other words, if the lot size is very large in time constant units, one sample of finite size suffices, but if the grab size is near zero, a number of samples must be taken. Note that here the sample can be obtained from one part of the object (e.g., when $G \neq 0$) or from several, equally spaced places (e.g., when $G = 0$).

When $T_x \to \infty$ the situation of a homogeneous object is approached. In this case it can be shown that σ_m, σ_μ, and $\sigma_{m\mu}$ all approach the value of σ_x; therefore σ_* will be 0 and one sample of any desired size suffices. In Figure 2.5 the relation of P, F, and n for a constant value of $\sigma_*/\sigma_x = 0.1$ is given. At low values of P both a small grab size F and a small number of grabs n are needed to obtain this value. This means that in highly autocorrelated lots only one small sample is needed to describe the lot with sufficient accuracy.

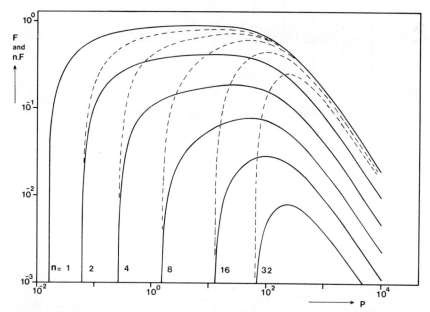

Figure 2.5. The relative sample size $F(—)$ and the relative gross sample size $nF(---)$ as a function of the relative lot size p for various number of samples ($\sigma_*/\sigma_x = 0.1$) (KA 78).

For medium lot sizes P, or values of T_x between 0 and ∞, the smallest sample needed to obtain a certain value is one made up from many small grabs. When fewer but larger grabs are taken, the size of the sample increases. For $F=0$, therefore, infinite small grabs, and a given value of n, σ_*/σ_x can be equal to 0.1 only for values of P larger than those indicated in Table 2.2. If the lot size P is less than these values, σ_*/σ_x is less than 0.1. For large values of P and σ_*/σ_x constant, $\log F$ is proportional to $\log P$.

In Figure 2.6 the relation between n, G, and σ_*/σ_x for some values of T_x and a fixed value of P is given. It is possible to reach lower values of σ_*/σ_x for medium values of T_x in two ways: by increasing the number of samples n or by increasing the grab size G. Over certain values of G increasing n is impossible because nG, the size of the gross sample, cannot exceed the lot size P without overlapping grabs G. For lower values of T_x an increase in sample size F causes a larger improvement in the precision than for higher

TABLE 2.2 Minimal Values of p to obtain $\sigma_*/\sigma_x = 0.1$ ($F=0$) (KA 78)

n	1	2	4	8	16	32	64	100
p	0.0151	0.0614	0.2640	1.508	12.64	64.34	342.0	∞

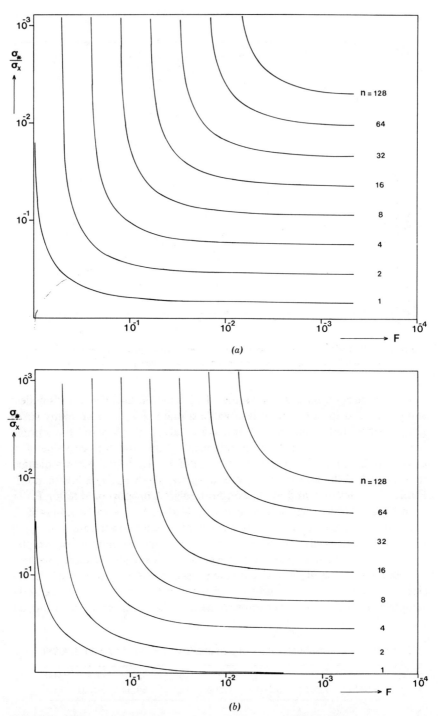

Figure 2.6. Relationship between σ_*/σ_x and the sample size G for a given lot size (P) and time constant (T_x) and for different numbers of samples (n) (KA 78). (a) $P=1000$, $T_x=1000$, $p=1$; (b) $P=1000$, $T_x=100$, $p=10$; (c) $P=1000$, $T_x=10$, $p=100$; (d) $P=1000$, $T_x=1$, $p=1000$.

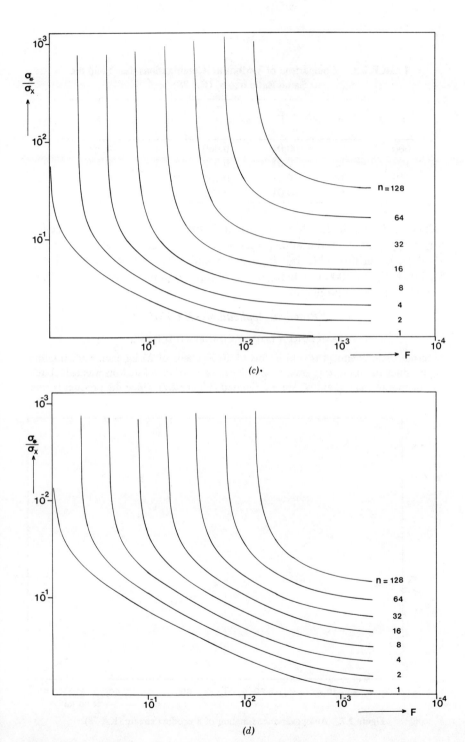

(c)·

(d)

Figure 2.6 (*continued*)

33

TABLE 2.3. Comparison of Various nF Combinations that Yield the Same Ratio σ_*/σ_x (KA 78)

p	n	F	$\dfrac{\sigma_*}{\sigma_x}$	n	F
1000	32	0.01	0.060	1	0.350
100	32	0.01	0.079	1	0.730
10	32	0.01	0.031	1	0.960
1	32	0.01	0.015	1	0.980

values of T_x. Table 2.3 shows F, the grab size G as fraction of P, needed for one grab to reach the same precision as is reached by 32 samples of size $F = 0.01$. From this table and Figures 2.5 and 2.6 it can be concluded that for lots with $P < 1000$ it is better to use a large number of small grabs than a small number of large grabs.

Practical Applications (KA 78)

SAMPLING A LOT OF CORRELATED ITEMS

A certain bulk chemical is sold in lots of 10,000 bags of 25 kg each. By sampling the product stream during production, the autocovariance function was calculated and the parameters σ_x and T_x were estimated (Figure 2.7). Since the product is sold

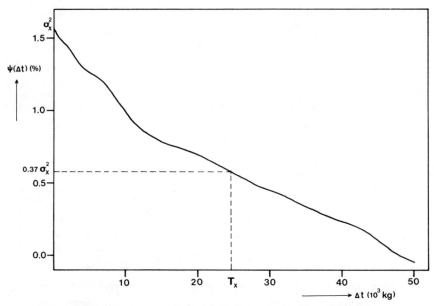

Figure 2.7. Autocovariance function of a product stream (KA 78).

TABLE 2.4. Size of Gross Sample as a Function of the Number of Samples for a Specified Value of $\sigma^*/\sigma_x = 0.08$ (KA 78)[a]

No. of Samples n	Sample Size F	Sample Size $G = PF$ Bags	Gross Sample nG Bags
1	0.91	9100	9100
10	0.045	450	4500
18	0.0014	14	252
19	0	1	19
20	0	1	20
40	0	1	40

[a] $\sigma_x = 1.25\%$; $\sigma_* = 0.1\%$; $T_x = 24,600$ kg $= 984$ bags; $P = 10,000$ bags; $p = 10,000/984 = 10.2$.

under the condition that the analysis of the whole lot will not deviate by more than 0.3% from the specified composition, a 95% confidence level is adopted. The standard deviation of the gross sample should not exceed 0.1%, accounting for the standard deviation of the method of analysis. The sampling scheme of Table 2.4 can be computed from eqs. 2.9–2.12. Note that a sample size G of less than one bag still requires one sampling action. It is clear that, in this case, sampling 19 bags is the method of choice. When it is assumed that the lot is not internally correlated, the data of Table 2.5 are valid; here more samples are needed. The incorporation of the time constant T_x in the calculations can decrease the number of samples from 150 to 19.

SAMPLING A RIVER

For estimation of the annual mean of a component in a river, a certain number of samples must be taken. If the internal correlation of the component is considered, the precision will be higher. Otherwise, an accepted precision can be obtained with fewer samples. For the concentration and load of some components of the Rhine

TABLE 2.5. Size of Gross Sample as a Function of the Number of Samples for a Specified Value of $\sigma_*/\sigma_x = 0.08$, Assuming $T_x = 1$ (Bag) (KA 78)[a]

No. of Samples n	Sample Size F	Sample Size $G = PF$ bags	Gross Sample nG Bags
1	0.03	300	300
10	0.0029	29	290
20	0.0014	14	280
50	0.00048	5	250
100	0.00014	2	200
150	0	1	150
200	0	1	200

[a] $\sigma_x = 1.25\%$; $\sigma_* = 0.1\%$; $P = 10,000$ bags; $p = 10,000/1 = 10,000$.

TABLE 2.6.   Parameters of the Rhine River for the Concentration and Load Values of Two Components. Estimated from Data Measured in the Period 1971–1975 (KA 78)

Variable	$\bar{x}$ (mg/l)	σ_x (mg/l)	σ_a (mg/l)	T_x (days)
NH_4^+				
Concentration	3.07	1.87	0.62	120
Load	4.91	2.07	0.93	50
NO_3^-				
Concentration	13.43	4.79	1.47	90
Load	24.22	11.71	2.67	30

River at Bimmen (West Germany) σ_x, σ_a, and T_x were estimated from the autocovariance function (Table 2.6).

The error of the determination of the mean value of a component during one year, Δ_*, depends on the values of σ_* and σ_a, the precision of the method of analysis. Assuming that each sample is followed by an analysis, the ultimate precision is $\sigma_L^2 = \sigma_*^2 + \sigma_a^2/n$. For a confidence level of 0.05, $\Delta_* = t_n\sigma_L$ where t_n is Student's-t for a confidence level of 0.05 and n determinations. The calculations are performed for two samples sizes, $G=0$ (in this case the samples are called momentary) and $G=A$, if the samples are integrated between the sampling actions. In the latter case, with $nG=P$, the value $\Delta_* = t_n\sigma_a/n^{1/2}$ is to be expected. For both

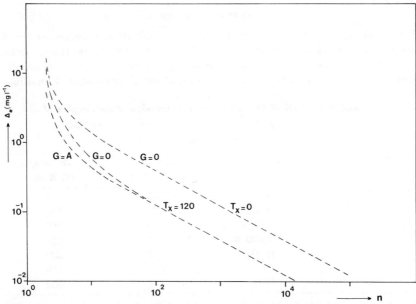

Figure 2.8. The error in the estimation of the annual average Δ_* as a function of the number of samples n (KA 78).

TABLE 2.7. Number of Samples n needed to Obtain a Certain Ratio $\Delta_* / \bar{x}$ in Estimating the Annual Average of a Variable Investigated in the Rhine River (KA 78)[a]

Variable	Ratio $\Delta_* / \bar{x}$			
	0.1	0.05	0.01	0.005
NH_4^+				
Concentration	24	73	1600	6300
Load	22	65	1400	5500
NO_3^-				
Concentration	12	29	480	1900
Load	19	43	510	1900

[a]$G = 0$; other data in Table 2.6.

cases the relationship between Δ_* and the sampling interval A or the number of samples n, calculated for the concentration of ammonia in the Rhine River, are shown in Figure 2.8. The same relationship is also shown for $G = 0$ neglecting of the autocorrelation ($T_x = 0$). If the number of samples n is larger than 130 the results for $G = 0$ and $G = A$ are practically identical. Furthermore, it can be shown that samples collected over 1 day ($G = 1$) barely give a better estimation of the annual average than the momentary samples ($G = 0$).

From this graph conclusions can be drawn about the sampling strategy. The best strategy (with the lowest value of Δ_*) is to take a sample collected over the whole period P, which should be analyzed afterwards by a very precise method. However, if a precise method is not available, or if the samples cannot be conserved over the whole period, more samples must be taken and analyzed. The number of samples needed to obtain a certain value of Δ_* can be deduced from Figure 2.8.

Frequently, a relative error $\Delta_* / \bar{x}$, where $\bar{x}$ is the mean value of the variable, is required. Table 2.7 shows how many samples are needed to obtain the ratio

TABLE 2.8. Number of Samples n Needed to Obtain a Certain Ratio $\Delta_* / \bar{x}$ in Estimating the Annual Average of a Variable Investigated in the Rhine River, if Autocorrelation is Neglected (KA 78)[a]

Variable	Ratio $\Delta_* / \bar{x}$			
	0.1	0.05	0.01	0.005
NH_4^+				
Concentration	160	640	16000	63000
Load	85	330	8200	33000
NO_3^-				
Concentration	56	220	5400	21000
Load	97	380	9500	38000

[a]$T_x = 0$; $G = 0$; other data in Table 2.6.

$\Delta_* / \bar{x} = 0.1$, 0.5, 0.01, or 0.005. These calculations were performed for the estimation of the annual average of the variables of Table 2.6. The values of n, assuming no internal correlation ($T_x = 0$), are given in Table 2.8.

2.5.2 Samples for Description in Detail

In contrast with the gross description of objects, where one parameter, the mean, suffices to describe the object, here the composition as function of time or space is wanted. Such a situation is met in metallurgical analysis, for example, when the compositions of the various phases are to be determined.

One may ask how many samples are needed to reconstruct the object in sufficient detail. Every sample must fulfill the condition that it resembles the object, that is, the part of the object that has to be reconstructed, as well as possible by the method of analysis and the details sought.

Heterogeneous Objects

If we assume that the object to be described consists of many objects, each homogeneous and randomly distributed, it is clear that the sample size must be smaller than or equal to that of the object. A restriction is that the method of analysis must be able to discriminate between the composition of the constituting objects. The number of samples cannot be estimated, for this is set by the problem: is the object to be known exactly in every detail, or will it be sufficient to know the structure of the object? In this case the number of samples is given by the rules of Section 2.5.1.

Internally Correlated Objects

When the object is internally correlated, no distinct boundaries exist. In this case the sampling theorem of Nyquist can be used. This tells us that the sampling frequency must be twice as high as the highest frequency that occurs in the object. This rule works very well in physics, where many phenomena are expressed in wave forms that can be converted more or less easily into the frequencies by Fourier analysis. In chemistry, however, this approach is more difficult (Section 4.3.2).

To know the composing frequencies, one must know the exact composition of the object with a very high resolution, in order to construct the wave form. In this case it is easier as a rule to estimate the time constant, for this parameter can be obtained in many ways, as given in Section 4.5.2.

2.5.3 Samples for Threshold Control

Processes are often monitored to report the day-to-day or hour-to-hour state of the process. The purpose of monitoring can be detection of a predetermined threshold value, therefore raising an alarm. There are numerous examples of such procedures in practice, such as detecting an explosion limit and seeking the maximum allowable concentration of a catalyst poison or of a pollutant in an effluent.

When the only function of the monitoring system is to raise an alarm, many samples are superfluous because, as a rule, the process value does not exceed the threshold levels. An optimal situation may be defined to be where the number of samples is such that a predetermined fraction of all threshold transgressions is detected. The selected value of this fraction depends on the costs (expressed in some measure, e.g., money) of effecting sampling and measuring, and the costs that accompany exceeding the threshold level.

Two ways to tackle this problem are described. The first is to adapt the sampling frequency to the last known process value: if this value is remote from the threshold level, the sampling frequency is low. If, however, the process value approaches the threshold level, the sampling frequency is enhanced.

The second way of attack is to fix the sampling frequency according to the mean probability of threshold transgressions. Which way is preferred depends on the method of sampling and the absolute values of the times between samplings. In some instances a rigid sampling scheme is required; in other instances, for example, in sampling a river with a large time constant, sampling may depend on the results of the last sample.

Variable Distance between Samples

When monitoring with the aim of detecting a transgression of an upper or lower threshold level, or both, the most reliable scheme is to measure continuously, for in this way all threshold transgressions are detected.

As a rule not as many analyses are required when not all transgressions have to be detected. When the cost of analysis exceeds the cost due to excessive process values, such a situation becomes interesting.

When the situation of a first-order stochastic process is assumed, which is often encountered, the frequency of sampling can be adapted to the process situation as described by Müskens (MU 78).

Let us assume that the process to be monitored has a time constant T_x, as follows from the autocorrelation (Section 4.5.2), a mean process value

μ_x, and a standard deviation of the process σ_x. The threshold level x_g can be chosen or will be dictated by the environment.

Future values can be predicted according to the equation

$$\hat{x}_{t+\tau} - \mu_x = \alpha^\tau(x_t - \mu_x) \qquad (2.13)$$

where $\hat{x}_{t+\tau}$ = predicted value of x, τ units of time after now (t), and
$\quad\quad x_t$ = process value at time t.

The variance of the prediction is expressed by

$$\hat{\sigma}_\tau^2 = \sigma_x^2(1 - \alpha^{2\tau}) \qquad (2.14)$$

For first-order stationary stochastic processes $\alpha = \exp(-1/T_x)$. Thus after each measurement x_t the probable future values of x, when predicted over a time span τ, are found in a region with the boundaries

$$\text{upper} = \hat{x}_{t+\tau} + N_p\hat{\sigma}_\tau = \alpha^\tau(x_t - \mu_x) + \mu_x + N_p\sigma_x(1 - \alpha^{2\tau})^{1/2} \qquad (2.15)$$

$$\text{lower} = \hat{x}_{t+\tau} - N_p\hat{\sigma}_\tau = \alpha^\tau(x_t - \mu_x) + \mu_x - N_p\sigma_x(1 - \alpha^{2\tau})^{1/2} \qquad (2.16)$$

N_p is the reliability factor that sets the probability p that the real process value $x_{t+\tau}$ will be within the reliability interval. This factor can be chosen.

As long as the upper limit of the reliability interval does not cross level x_g (Figure 2.9) the probability that the threshold level will be exceeded is less than $1-p$ and sampling is not necessary. However, as soon as the upper level reaches x_g, sampling should be started. From the result thus obtained the time for the next sampling may be derived in the same way. By application of this procedure, which is based on the most recent information, the sampling frequency will be lowered automatically if the process values are remote from the threshold x_g. The frequency increases when the process values approach x_g. The prediction period τ, during which time no samples have to be taken, can be derived by fixing

$$x_g = \text{upper (or lower) limit}$$

Then

$$\tau = -T_x \ln \frac{Gy_t + N_p\left[y_t^2 - a(G^2 - N_p^2)\right]^{1/2}}{y_t^2 + aN_p^2} \qquad (2.17)$$

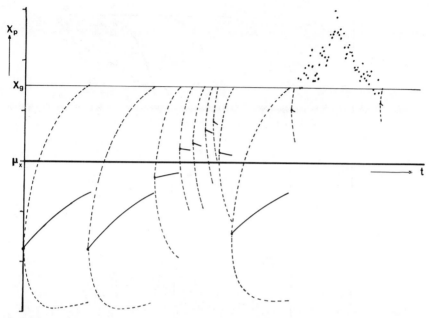

Figure 2.9. Operation of the monitoring system (MU 78). (...) measurements; (—) prediction; (---) 95% reliability interval of the prediction.

where $G=(x_g-\mu_x)/\sigma_x$
$y_t=(x_t-\mu_x)/\sigma_x$
$a=(\sigma_x^2+\sigma_a^2)/\sigma_x^2$
$\sigma_a^2=$ variance of method of analysis

Note that N_p must be chosen greater than G.

The sampling frequency is fixed by the process parameters, the threshold value, and the reliability factor. Because the process parameters are inherent to the process and the threshold level is set by the environmental conditions, the only degree of freedom is the reliability factor N_p. The higher N_p, the lower the probability of undetected threshold transgressions, but the higher the sampling frequency. The optimal value of the reliability factor can be determined by balancing the costs of the analysis against the costs caused by undetected threshold transgressions. The mean frequency of sampling $\bar{\tau}^{-1}=\mu_a^{-1}$ can be calculated (MU 78) and is plotted in Figure 2.10 as a function of the reliability factor N_p for some time constants.

The sampling frequency increases with $N(p)$. This increase is more rapid for lower time constants. For a time constant equal 0, the sampling frequency is 0 for $N_p<(x_g-\mu_x)/\sigma_x$ and is 1 for $N_p>(x_g-\mu_x)/\sigma_x$. If, for example, the threshold value $x_g=\mu_x+\sigma_x$ (16% of the process values larger

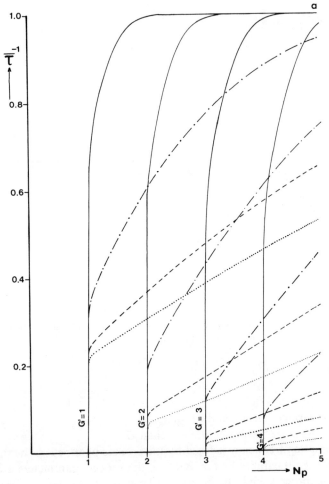

Figure 2.10. Relationship between the mean frequency of analysis $\bar{\tau}^{-1}$ and the reliability factor N_p for four threshold values (1, 2, 3, and 4) (MU 78). (—) $T_x = 1$; $(- \cdot -)$ $T_x = 10$; (---) $T_x = 50$; (...) $T_x = 100$. (a) One-sided problem; (b) two-sided problem.

than x_g), the time constant $T_x = 100$, and the reliability factor $N_p = 2$, the mean sampling frequency is 0.4; thus a decrease of 60% in the number of samples occurs compared with sampling with a frequency $1/\tau$.

The average probability p'_u that a threshold transgression occurs without having been detected can be calculated (MU 78). $p_u = 1 - p'_u$ quantifies the probability of a threshold transgression, which is detected. Indeed, the higher p_u, the more reliable the monitoring system.

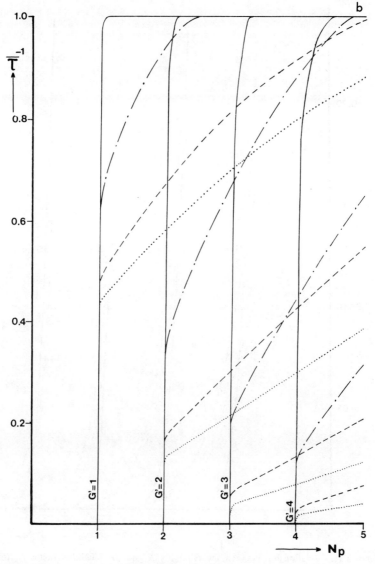

Figure 2.10. *Continued*

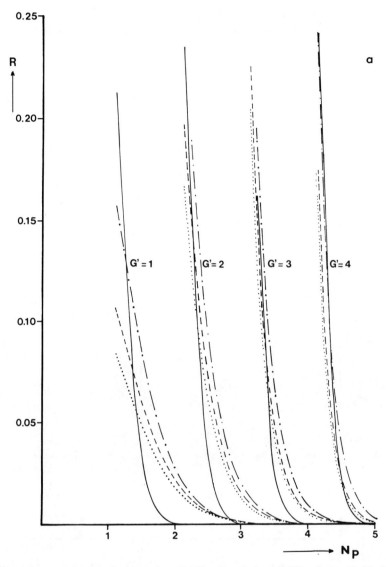

Figure 2.11. Relationship between the reliability R and the reliability factor N_p for four threshold values (1, 2, 3, and 4) (MU 78). (—) $T_x = 1$; (– ·–) $T_x = 10$; (---) $T_x = 50$; (...) $T_x = 100$. (*a*) One-sided problem; (*b*) two-sided problem.

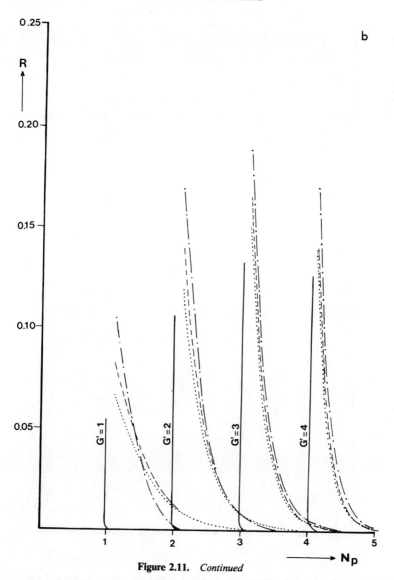

Figure 2.11. *Continued*

In Figure 2.11 the relationship between the reliability factor N_p and the reliability $R = (1 - p_u)/(1 - p_g)$ is shown, R being the fraction of undetected transgressions of all possible transgressions. If this ratio equals 1, no transgressions are detected. If, however, the ratio equals 0, all transgressions are detected. This can be achieved only by sampling all process values with a frequency $1/\tau$.

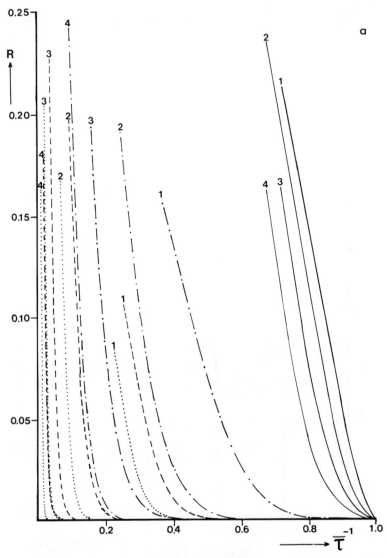

Figure 2.12. Relationship between the reliability R and the mean frequency of analysis $\bar{\tau}^{-1}$ for four threshold values (1, 2, 3, and 4) (MU 78). (—) $T_x = 1$; (— · —) $T_x = 10$; (- - -) $T_x = 50$; (...) $T_x = 100$. (a) One -sided problem; (b) two-sided problem.

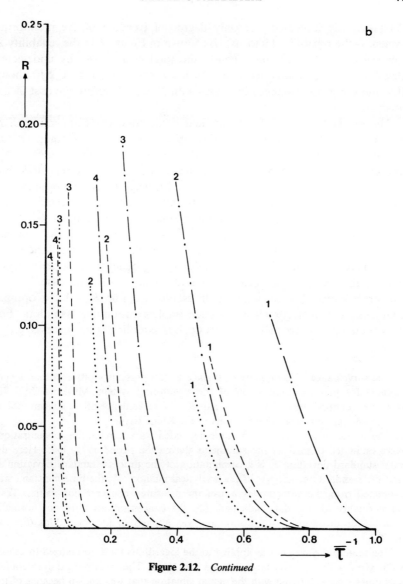

Figure 2.12. *Continued*

From Figure 2.12 it can be seen that a reliability R can be achieved when

$$N_p > G + 0.5, \qquad R > 90\%$$

$$N_p > G + 1.0, \qquad R > 98\%$$

$$N_p > G + 1.5, \qquad R > 99\%$$

As previously indicated, the only degree of freedom of the monitoring system is the reliability factor N_p. As shown in Figure 2.11 the reliability R increases with N_p and consequently the total cost caused by undetected threshold transgressions decrease. On the other hand, Figure 2.10 shows that the sampling frequency increases with N_p and therefore the cost of the analysis increase.

The optimal value of N_p is obtained if the total costs are minimal. A rather elaborate method of finding this optimum is to compare the frequency of analysis with the reliability for corresponding values of N_p, and calculate the costs of both components. This should be repeated until the total minimal costs are found. A helpful way to find the optimum as quickly as possible is the simplex optimization method (see Section 4.8).

A less elaborate method to find the optimum is by plotting the reliability as a function of the sampling frequency (Figure 2.12). The optimal frequency and its reliability are found from the coordinates where the curve has a slope equal to $-(1-p_g)K_b/K_a$, K_a being the cost per analysis and K_b the cost per undetected threshold transgression. The term $1-p_g$ is the probability of exceeding the threshold x_g. For $K_a=0$ the optimal frequency equals 1, which in practice implies continuous sampling. For $K_b=0$ the optimal frequency equals 0; thus sampling is not necessary.

Example

The surveillance of the quality of surface waters presents a problem that is very suitable for practical application of the monitoring system. Müskens (MU 78) described an application in this field, namely, the surveillance of the concentrations of ammonium, nitrite, and nitrate ions in the Rhine River.

The statistical properties of the time-dependent variations in the concentrations were estimated from daily measurements during the period 1971–1975. Here the total standard deviation s_y is a combination of the process standard deviation s_x and the standard deviation s_z of the analytical method. T_x, the time constant, was estimated from the autocorrelogram of the measured concentration values. Tests on normality for the data indicated that the concentrations were not normally distributed; therefore, a transformation was applied to the logarithms (Section 4.1.2).

The monitoring system was applied to the logarithms for a normalized threshold value $G=2$ with reliability factors $N_p=3$ and $N_p=4$. The predicted mean interanalysis times were compared with the actual situation that was known because of the daily analysis. These values agreed very well. The mean interanalysis time for ammonium ion was

$$G=2 \quad N_p=3 \quad 4.8 \text{ days}$$

$$N_p=4 \quad 3.1 \text{ days}$$

for nitrite ion

$$G=2 \quad N_p=3 \quad 3.0 \text{ days}$$

$$N_p=4 \quad 2.0 \text{ days}$$

and for nitrate ion

$$G=2 \quad N_p=3 \quad 2.7 \text{ days}$$

$$N_p=4 \quad 1.8 \text{ days}$$

Because the accepted frequency of analysis was 1/day, it was concluded that considerable savings were possible if the proposed monitoring system was adhered to.

Fixed Distance between Samples

When a varying sampling frequency is not appropriate, a fixed frequency can be chosen in such a way that the sum of control costs and costs caused by exceeding the threshold is a minimum (KO 71). The mean time between transgressions of the threshold limit, or the mean undisturbed run length T_r, can be determined by counting the number of transgressions m during a time T and defining $T_r = T/m$. The aim is to detect transgressions of x_g, but because the only way to detect all transgressions is to sample continuously, an optimal value of the sampling frequency is sought.

Let us suppose the mean time between samplings is T_a (a sampling frequency of $1/T_a$) and the mean delay time (time required for sampling and analysis) is T_d (Figure 2.13). Because now the mean time between

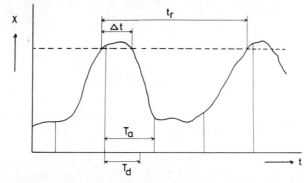

Figure 2.13. Process with undisturbed run length t_r, transgression length Δt, distance between samples T_a, and delay time of analysis T_d (KO 71).

transgression detections is $\frac{1}{2}T_a$ and the transgression is not under control before action has been undertaken, the mean time of transgression ΔT is $\frac{1}{2}T_a + T_d$. (We assume that the effect of the control action is immediate, for very extreme control actions are possible, such as shutting off the endangering process stream.) The fraction of time the process suffers transgression is

$$\frac{\Delta T}{T_r} = \frac{\frac{1}{2}T_a + T_d}{T_r} \qquad \left(\frac{\Delta T}{T_r} \ll 1\right) \tag{2.18}$$

If the cost of transgression (losses of the production process, fines for elevated emissions, and costs of extra actions to prevent damage) is C_p ($/hr), the cost due to off-limit situations is

$$C_L = \frac{\left(\frac{1}{2}T_a + T_d\right)}{T_r} C_p \tag{2.19}$$

If the sampling cost (including sampling and analysis) is c_a ($/sample) the hourly cost is $C_a = c_a / T_a$. The total cost of control ($/hr) is the sum of C_a and C_L, or

$$C_t = \frac{C_a}{T_a} + \frac{\left(\frac{1}{2}T_a + T_d\right)}{T_r} C_p \tag{2.20}$$

For given or estimated values of c_a, C_p, T_d, and T_r, C_t is a function of T_a. This implies a minimum value of C_t when

$$(T_a)_{\text{opt}} = \left(\frac{2C_a T_r}{C_p}\right)^{1/2} \tag{2.21}$$

and

$$(C_t)_{\text{min}} = \frac{2C_a}{(T_a)_{\text{opt}}} \left[\frac{T_d}{(T_a)_{\text{opt}}} + 1\right] \tag{2.22}$$

As a rule estimates of C_p are difficult to obtain.

For use in practice it must be taken into account that $(T_a)_{\text{opt}}$ is reduced to "easy" values, for a sampling time of 2.735 hours, for example, would be very impracticable. Because plant laboratories generally operate at schedules of $\frac{1}{2}$, 1, 2, 4, or 8 hours, following a system of three shifts a day, an effective sampling scheme

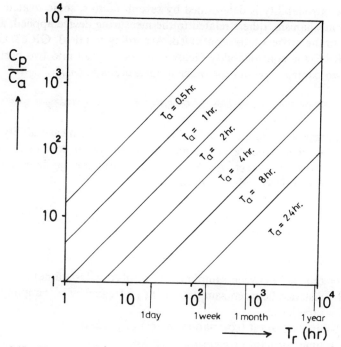

Figure 2.14. Nomogram for estimation of the optimum sampling frequency (KO 71).

must follow this schedule. A graph (Figure 2.14) can be produced that is based on these assumptions and uses only rough approximations of C_p/C_a.

2.5.4 Samples for Control

Parameters for Control (see also Section 3.2.5)

Because the purpose of control is to diminish the variance of the process, the parameter that describes the success or the maximum available success of controlling action is based on diminution of the variance. The controllability factor r is defined as

$$r^2 = \frac{\sigma_x^2 - \sigma_{min}^2}{\sigma_x^2} = 1 - \left(\frac{\sigma_{min}}{\sigma_x}\right)^2 \tag{2.23}$$

In this equation σ_x^2 = variance of process deviations without control and σ_{min}^2 = variance of process deviations after control. A factor $r = 1$ means that the variations in the process are fully eliminated and control is 100% successful. A factor $r = 0$ means that no diminution at all has occurred.

The controllability is determined by several factors, some related to the control mechanism, others related to the measuring device applied, and all related to the process to be controlled. According to GI 63, GR 63, GR 65, and GR 66, the controllability factor r can be divided into two parts, one related to control (r_p) and the other to measurement (m).

$$r = r_p m \qquad (2.24)$$

For the purpose of sampling only m is of interest. The measurability factor m describes the influence of the measuring device on control. When the control action is perfect, $r_p = 1$; in such a situation m determines the maximum obtainable value of r.

$$m = \exp\left(\frac{-(D + A/2 + G/3)}{T_x} \right)\left[1 - \frac{\sigma_a \sqrt{T_e}}{\sigma_x \sqrt{T_x}} \right] \qquad (2.25)$$

with $D =$ time lag between sampling and result (analysis time)
$A =$ distance between samples $= 1/f$ ($f =$ frequency of sampling)
$G =$ sample size
$T_x =$ time constant (correlation factor) of process
$T_e =$ time constant of measuring device
$\sigma_a^2 =$ variance of measuring method
$\sigma_x^2 =$ variance of process
This equation can be rewritten as

$$m = m_D m_A m_G m_N \qquad (2.26)$$

in which

$$m_D = \exp\left(-\frac{D}{T_x} \right) = \exp(-d) \qquad (2.27)$$

$$m_A = \exp\left(-\frac{A}{2T_x} \right) = \exp\left(-\frac{a}{2} \right) \qquad (2.28)$$

$$m_G = \exp\left(-\frac{G}{3T_x} \right) = \exp\left(-\frac{g}{3} \right) \qquad (2.29)$$

$$m_N = 1 - \frac{\sigma_a \sqrt{T_e}}{\sigma_x \sqrt{T_x}} = 1 - s_a \sqrt{t_e} \qquad (2.30)$$

and $D/T_x = d$, $A/T_x = a$, $G/T_x = g$, $T_e/T_x = t_e$, and $\sigma_a/\sigma_x = s_a$.

In order to describe the influence of sampling it can be stated that

$$r = (m_D m_N) m_S r_P \qquad (2.31)$$

m_S being the factor describing the sampling action

$$m_S = m_A m_G \qquad (2.32)$$

Distance between Samples

From eq. 2.26 it follows that the maximum obtainable controllability factor will never exceed the smallest of the composing factors. This implies that all factors should be considered in order to eliminate the restricting one. It also means that a trade-off is possible between high and low values of the various factors. If, for example, m_N, the factor influenced by the reproducibility of the measuring device or the method of analysis, is high and m_A is low, the frequency of sampling can be increased, causing m_A to increase. The reproducibility of the analysis can be decreased—which may result in a less expensive means of analysis—in order to cope with the higher rate of sampling if the product $m_A m_N$ is higher than before.

From $m_A = \exp(-a)$ it follows that decreasing a, the distance between samples or the reciprocal of the sampling frequency, causes m_A to increase. The higher the sampling rate, the better the possibility of controlling the process. This does not mean, however, that the highest obtainable frequency is the best, for sampling costs rise about linearly with $1/a$. Moreover, the optimal value of a depends on balancing the costs of analysis against costs of an eventual process failure, as is shown below.

From $m_G = \exp(-g/3)$ it is clear that decreasing g, the sample size, increases m_G. The result is that, when sampling for control, it is always better to use samples that have the minimum size required by analysis. This is contrary to the situation when describing a lot. A general statement can be made: never use integrating devices that collect samples during the time between analyses. If, for example, $g = a$, the sample is collected during the time between two consecutive samplings. Here

$$m = \exp\left[-\left(\frac{a}{2} + \frac{g}{3}\right)\right] = \exp\left[-\left(\frac{5a}{6}\right)\right] \qquad (2.33)$$

and high values of a result in low values of m. A special case is encountered when the time constant of the instrument, for example, an automatic analyzer, sets the sampling rate. Here $T_e = a$, so a third factor is influenced by sampling:

$$m_S = m_A m_G m_N$$

$$= \exp\left[-\left(\frac{a}{2}+\frac{g}{3}\right)\right]\left(1-s_a\sqrt{a}\,\right) \tag{2.34}$$

Practical Applications

The assumption that production costs rise exponentially with deviations from the set point yields

$$K' = [1+\alpha(\exp|s|-1)]K \tag{2.35}$$

with α = factor affecting the price of deviations
$\quad K$ = cost of product when produced at specification level
$\quad s$ = dimensionless deviation from specification level expressed in standard deviations

The probability of producing at specification is

$$P = \frac{2}{\sqrt{2\pi}}\exp\left(-\tfrac{1}{2}s^2\right) \tag{2.36}$$

The cost caused by deviations from the set point thus is

$$k = K\left[1+\frac{\alpha}{\sqrt{2\pi}}\int_{-\infty}^{\infty}(\exp|s|-1)\exp\left(-\tfrac{1}{2}s^2\right)ds\right] \tag{2.37}$$

By control the deviations of the process are diminished and the expression becomes

$$k' = K\left[1+\frac{\alpha}{\sqrt{2\pi}}\int_{-\infty}^{\infty}(\exp|s|-1)\exp\left(\frac{-s^2}{2(1-m^2)}\right)ds\right] \tag{2.38}$$

The influence of control on process costs can be expressed by

$$c = \frac{k'}{k} = \frac{\int_{-\infty}^{\infty}(\exp|s|-1)\exp\left[-s^2/2(1-m^2)\right]ds}{\int_{-\infty}^{\infty}(\exp|s|-1)\exp\left(-\tfrac{1}{2}s^2\right)ds} \tag{2.39}$$

The assumption that production costs rise linearly yields

$$K' = (1+\alpha|s|)K \tag{2.40}$$

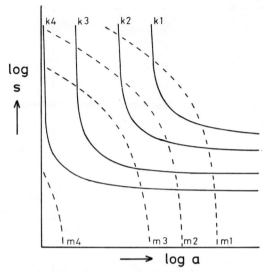

Figure 2.15. Cost and measurability as a function of precision (S) and analysis frequency (a^{-1}) [*Chem. Weekbl.* **1978**, 12 (May)]. Cost K_1 100 arbitrary units; K_2 200 arbitrary units; K_3 500 arbitrary units; K_4 1000 arbitrary units. Measurability m_1 0.8, m_2 0.9, m_3 0.95, m_4 0.99.

and

$$c = \frac{\int_{-\infty}^{\infty} |s| \exp\left[-s^2/2(1-m^2) \right] ds}{\int_{-\infty}^{\infty} |s| \exp(-s^2/2) \, ds} \tag{2.41}$$

Because m is a function of a, the distance between samples, it is possible to use a second axis giving a, as shown in Figure 2.15. It is clear that the knowledge of cost factors K, together with the cost of sampling as a function of a, provide the optimal situation because the sum of yield (by increasing m) and costs (caused by increased sampling rate) becomes maximal.

In order to provide some feeling for the cost factors under consideration one is referred to Leemans (LE 71), who published some data on the measurability and costs of an industrial process (Table 3.3 and Figure 2.16). The relation between m and cost is plotted in Figure 2.17.

If this figure is redrawn, another representation shown in Figure 2.18 clearly demonstrates that there exists a restricted area that sets boundaries to the variability of a. Low values of a result in excessively high costs of analysis, unless the methods are fully automated. Large values of a are reflected in an undesired rise of the process costs. Note that Leemans uses a somewhat refined model for process costs. In his model he assumes that the margin left for process fluctuations in the guarantee conditions could be used if m is very high. This assumption implies that the extra process costs become negative at $m = 0.92$, as indicated in the figure.

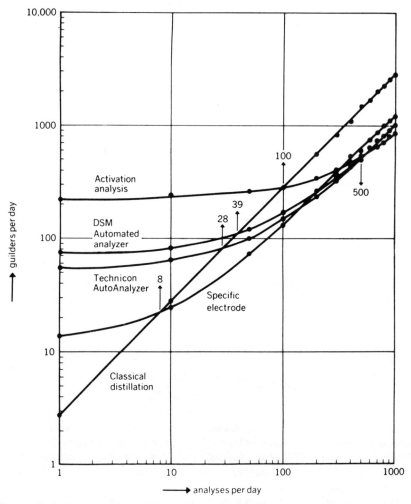

Figure 2.16. Costs of some off-line analytical techniques for the analysis of inorganic nitrogen (1968) (LE 71). Reprinted with permission from *Anal. Chem.* **43**(11), 46A (1971). Copyright by the American Chemical Society.

2.5.5   Samples for Special Purposes

One-Shot Sampling

In many instances it is not possible to use any of the sampling strategies that are mentioned in the preceding sections. Such a situation is met, for instance, when sampling the moon, the remains of a burnt building, or

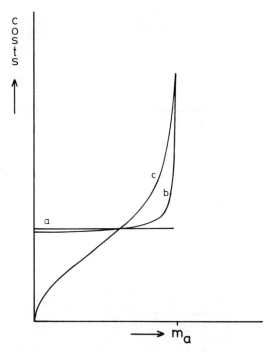

Figure 2.17. Cost of analysis as a function of measurability caused by analysis frequency [*Chem. Weekbl.* **1978**, 12 (May)]. (*a*) Capital-intensive; (*b*) intermediate; (*c*) labor-intensive.

evidence at the scene of a crime. This is the so-called one-shot sampling. Only one sample can be obtained, of a size that cannot be related to the object to be described. Sampling the moon cannot be done by an astronaut or an automatic moon sampler (EN 73) in the way described in Section 2.5.1 or 2.5.2. When describing the place to be sampled there is very little known about the distribution of the components, their composition, or the correlation, for example, so it is not possible to state the size or number of grabs that must be taken to constitute a real sample, representative of the entire moon.

These circumstances do not occur solely on the moon, but also when, for example, collecting geological samples on earth or taking samples in shops for the food inspection department. To overcome these problems it is possible to work the other way round. The size of the grab is determined by maximum weight, volume, or cost. The grab is stated to be a sample; when the sample has been analyzed, the parameters found are used to describe the lot size. In some situations the lot size is barely larger than the sample, such as when sampling parts of a large lot, for example, bread in

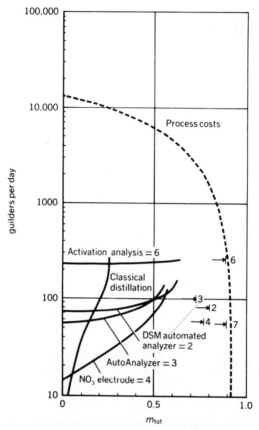

Figure 2.18. Process costs due to process fluctuations (dashed line) and analysis costs (solid line) as a function of analytical information. The arrows refer to the fully automated techniques (LE 71). Reprinted with permission from *Anal. Chem.* **43**(11), 44A (1971). Copyright by the American Chemical Society.

shops. Each loaf of bread sampled at the factory can be considered as a part of the bread making process, and will show correlation with loaves produced at the same time or found in the same oven. When the loaves are distributed in town, however, nothing can be said about the autocorrelation. The sample of one loaf equals a lot of one loaf.

When sampling the moon, the sample brought back to earth should be allocated to a type of rock known to exist in a known area. Here the description of the lot by the sample is part of the job. Another example (BU 73) of one-shot sampling is examination of evidence at the scene of an accident. The piece of glass found near the victim is known to be part of a lot, from the windshield of a car. Here, too, the problem is to allocate the sample to a lot. There are two ways of doing so. One is to allocate the sample to an unspecified lot; for example, this glass is used in windshields of manufacturer X. The other method is to allocate it to a specified lot; for example, this is glass from the broken windshield of car Y.

In the case of one-shot sampling, care must be taken not to overestimate the evidence provided by the sample. As is shown in the preceding chapters, many properties of the lot or process must be known in order to ascertain the size of a sample that is representative of the lot. Only when these properties can be obtained from the sample itself it is justified to describe the lot using the data the sample yields.

Selected Sampling

Although it is good practice to define the sample as representing the lot or the process, there are situations in which this rule should not be applied. Two possibilities are described.

1. *Tracing Deviations from Normal Situation.* This occurs when faults in material must be found, mistakes, or forgeries, for example. Now the purpose is not to describe the whole process or lot with one sample, but to prove that part of the process or lot is deviating. In order to relate the results derived from the sample to the part considered, this part can be taken as a partial lot and the normal rules can be applied. Note that a part of a process, smaller than that needed for an ergodic part, by definition is a lot (see Section 2.5.1). Here the rules for lots are to be used.

2. *When Deviations of the Normal Situation are Suspected.* In many cases deviations are suspected, for example, when malfunction of the production process is possible or when forgeries are suspected. The recommendation in such a situation is to forget everything you know about sampling and use your common sense and experience. As soon as the deviation has been found, return to the scientific practice and use the rules of part 1.

2.6 HANDLING OF SAMPLES

The foregoing parts have been devoted to the theoretical requirements of sampling. In practice, however, the value of a sample is determined not only by the prerequisites of time and size, but also by the handling of the sample. This means that the sample must be taken, homogenized, reduced in size, packed, labeled, and protected against contamination and decomposition. Insight into the foregoing theory and common sense will help in improving the quality of sampling. Although sampling hardly can be learned without actual practice, a host of books and papers on sampling practice is available (AS 51, BO 52, AS 54, BE 54, AS 56, TO 59, WA 59, WE 63, AN 77). Apart from the literature mentioned, some papers deal with sampling equipment (BE 45, SP 65, KA 66). Directions for sampling specific products can be found in PE 38, AM 56, AT 56, TO 59, WA 59, AO 60, AS 73, EB 75, JA 76, and WA 76.

2.6.1 Homogenizing

Often the sample is composed of grabs. Because these grabs by definition are not representative of the object as a whole, the sample is inhomogeneous. If the sample is to be analyzed as a whole, this inhomogeneity is not important to the sampler, but usually only a part of the sample is used for analysis. Other parts are retained for duplicating the analysis, for counter-samples, or for legal purposes. This implies that the sample will be sampled again, each sample subdivided into subsamples. Because the sample itself can be of the minimum size required, it is often not possible to take a subsample without further action. As follows from the theory of sampling inhomogeneous objects, the sample size is dictated by the number of composing parts or the autocorrelation of the object. There are two methods of obtaining a good sample that can be subdivided: decreasing the autocorrelation, since a low value of the correlation constant permits small samples, and decreasing the particle size, for instance, by crushing or grinding. Here as a rule homogenization is accomplished automatically. Homogenizing is also possible during the sample size reduction.

Small Samples

Small gaseous samples do not need homogenizing because usually the sample container is filled with a turbulent gas stream and this is sufficient to mix the contents. Furthermore, diffusion in gaseous samples is quite effective at room temperature, so storing the sample for some time suffices in situations where the homogeneity is questioned. If the sample container has been filled with a laminar gas stream or if for another reason the sample seems liable to inhomogeneity, it is possible to use mixing aids such as glass beads, metal balls, or small sheets of metal. When the sample container is revolved, these aids mix the gas quite easily.

Small liquid samples do not pose any difficulties whatsoever. Partly filled containers can be shaken by hand, and totally filled containers can be mixed on a mechanical roller. Sludges and fluids with suspended particles are by nature very difficult to homogenize. Here special instruments are mandatory. These instruments are like those used for subdividing granular material.

Small solid samples pose the most difficulties. Powdered or fine granular material can be mixed similarly to fluids, provided the granules show a narrow particle size distribution. When fines are present, however, these tend to settle or agglomerate. The sample should be crushed or ground until a uniform particle size has been obtained. This method is inapplicable when the particle size distribution of the object is the aim of investigation. Coarse samples of brittle material can be ground as well. It is recommended that a sieve be used during grinding, to separate the unground material from the already ground particles. The coarse fraction is recirculated until all material passes through the sieve. The size of the sieve is set by the rules for obtaining a good subsample and by the properties of the particles as far as segregation is concerned. This method saves time and, more importantly, does not bring the sample into contact with the material of the mill or mortar more than necessary.

Besides mortars of china, agate, silicon carbide, tungsten carbide, or diamond, hammer or crushing mills are recommended. Ball mills should be used only when the analysis requires very fine samples.

Samples of metals are often in the form of turnings or shavings. Here homogenizing is very difficult because crushing or grinding is often impossible, except for the more brittle metals. The method of sampling should be modified to enable mixing of the sample before subsampling.

Materials that resist grinding or crushing, such as fibers or plastic materials, must be treated another way. Subsampling a sample of fibers can be very cumbersome. Unless specialized instruments, such as combs that make a fleece, are at hand, homogenizing must be done by "counting out." Every fiber or small bundle of fibers is assigned to a subsample, taking each subsample in turn and repeating this action until the whole sample has been treated and all fibers are divided between the subsamples.

Plastic material, like many polymers and foods, may often be kneaded on kneading rollers. However, when mechanical properties or chemical properties that are affected by kneading are involved, homogenizing is not possible and subsamples should be obtained in the form of multiple samples.

When the sample has to be used for description in detail, subsampling can pose problems. With solids it is difficult to state whether the fine structure is homogeneous throughout the sample. Here the best way is to estimate the correlation constant of the phenomenon sought and use a chunk of the material of a size that fulfills the requirements of a sample. Of course this can hardly be called homogenizing. In fact it is selecting subsamples in such a way that each subsample resembles a sample as far as possible.

Larger Samples

When sampling coarse or very inhomogeneous objects, the sample size can be considerable if the sample has been taken according to the rules of Section 2.5. A sample of coke with a mean particle weight of 1 kg can easily be several tons. When air has been sampled, for instance, in assessing pollution, or when large lots of fluids or slurries are sampled, sample sizes of thousands of liters or kilograms are not uncommon. Of course the same rules as applied to small samples have to be used here, but the execution is often quite different.

When voluminous gas samples must be collected, there are two possibilities. Samples under atmospheric pressure are often collected in inflatable bags. The content of a bag is very homogeneous because temperature differences, handling, and diffusion tend to cancel any inhomogeneity. When the sample is collected under pressure, whether this is done by compressing the sample or sampling pressurized gas, homogenizing is difficult. The best way is to put some mixing aids in the pressurized vessel and homogenize the contents of the pressurized vessel by rolling. (Another way is to carry the container some hours in a bad car on a bad road. Commuter cars are often more valuable for samples than for people.) As a rule, however, when large volumes of gas must be sampled, the sample is treated on the sampling site to collect the substance to be determined by extraction: drying filters, absorbing fluids, mechanical filters, or cold traps can be used. The contents of these aids can be treated like samples of fluid or particulate character.

Homogenizing large quantities of fluids is usually not difficult if the fluid is monophasic. If more phases are present, for example, two or more layers, emulsions, suspensions, or slurries, homogenizing can be a hard task. Usually efficient mixing and collecting the sample while mixing suffice. However, for fluids with phases that differ greatly in density, "centrifugal" effects are a nuisance; in this case shaking is more effective than stirring.

Fluids that have to be kept separated from the atmosphere, for example, condensed gases, oxidizable fluids, fluids liable to decompose with water, or fluids that have to be kept sterile, are always kept in closed vessels. If there is a headspace, and for safety reasons there should be, shaking suffices.

Homogenizing large quantities of solids is a difficult task. Usually this is accomplished by reducing the sample size and then grinding the entire sample.

For medium-sized samples (tens to hundreds of kilograms), various mixing devices that can shake the container in many complicated patterns can be useful, though this is no solution for samples of wide particle size distribution. A good method, though liable to bad results if not executed properly, is to collect chunks by hand (e.g., coal, ore, minerals, food) in such a way that each particle size is represented in the right proportion. This method can be excellent if the person who collects the subsamples is aware of what he is doing: subsampling a sample in such a way that each subsample is like the gross sample.

2.6.2 Reduction of the Sample

As has been stated in the preceding section, subsampling as a means to reduce the size of the gross sample to more manageable sizes, or to obtain more samples for different purposes, must be preceded by homogenizing (EN 70).

Subsampling gases and liquids, when properly mixed, does not pose many difficulties. Subsampling solids, large samples, or samples with wide particle size distribution as a rule requires more sophisticated methods.

An old and widely known method is quartering. The sample is put in a heap, roughly mixed, and subdivided into four quarters. Two opposite quarters are eliminated, the remaining two are mixed again, quartered, and so forth, until the size of the remaining sample is manageable by other methods. However, the method of quartering is often used into the gram region.

A handy instrument for subdividing is the riffle, a slanted plane that contains separated partitions (Figure 2.19). (WO 79) By collecting only half of the alleys, the sample size is halved. Here the sample is not

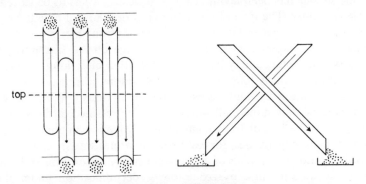

Figure 2.19. Riffle sampler.

homogenized. Many subsamples with grab size $G = \frac{1}{2} A$ are collected and yield a gross sample. The procedure can be repeated by recirculating the collected (half) sample. During the "reriffling" a certain amount of mixing occurs. One instrument reduces the sample to about one-eighth to one-sixteenth of its original volume (three to four cycles). Loading the riffle with a part of the sample, eliminating the odd loads, allows extended reduction. Use of a smaller riffle allows further reduction in the size of the sample. When the sample has been reduced in size sufficiently and is mixed by riffling, the riffle can be used to collect a number of subsamples.

Another mechanical means for sample reduction is the carousel. From a large hopper the sample is fed onto a string of smaller containers, mounted on the circumference of a disk. Each container in turn receives a small part of the sample. The disk is revolved as long as sample flows from the hopper. The number of subsample containers determines the sample size reduction. Of course other realizations of the same principle are possible, for example, a sample hopper revolving over the subsample containers. The carousel method can be used only for samples with a maximum size of a few kilograms. For larger samples the method has been incorporated into the sampling machine: the sample is fed onto a revolving disk. After each revolution a small section is swept into a sample collector; the remaining part is swept back into the product stream.

Note that from the preceding theory it follows that this type of sample reduction apparatus can be used only when some knowledge is available on the internal correlation of the samples. It is customary to homogenize the sample before reduction in order to make $p = P/T_x$ larger by reducing T_x (Section 2.5.1).

2.6.3 Sample Storage and Transport

After the sample has been taken or reduced in size, it has to be stored in a suitable container. The type of container used depends on how the sample is to be handled. In general it can be said that the container must be sturdy to protect the sample from mechanical damage during transport to the laboratory; the distance may be short, but is often quite long. The sample may be kept for a fairly long time, and furthermore, it should be protected from contamination by, and losses to, the environment. When the water content is to be determined, all water should be retained in the sample. A special case is the loss of trace components. Organic components that are present in minute quantities, for example, in natural waters, can be lost by adsorbtion onto the surface of the container, by absorbtion into rubber stoppers or sealing rings, by oxidation or decomposition under the influence of light, or by evaporation and diffusion through container walls.

Trace metals can be lost by adsorption on the wall of the container, and mercury can be lost by evaporation of the metal or, more likely, its organic complexes. A special case is the loss of sample components by biological processes: rodents and insects can affect samples of food, and bacteria and fungi may alter not only food but also many organic components and even inorganic salts.

Another cause of sample alteration can be human interference. In particular, when samples are used for trade or legal purposes it is important that no changes whatsoever can be caused by humans.

This leaves only a few types of suitable containers. When dry solids are to be packed, and their water content is not important, paper bags, plastic bags, or a combination of these are used. For trade or legal samples provision should be made for the container to be sealable.

Glass is one of the most suitable materials, either as glass jars with a screw cap or flasks. Most solids can be kept in glass containers for a sufficient length of time. Air and water can be excluded; fines and water are retained. Most solids (except some strong alkalines) do not affect glass.

Most liquids can be kept in glass also. As mentioned, special care is required when the sample contains trace components that should be determined. Since the adsorption of metals is probably caused by ion-exchange effects, a low pH of the sample seems to be a prerequisite (e.g., KI 74, ST 73, RY 72, RO 68, RO 73). Care must be taken that no components from the glass will contaminate the sample. In general, traces of sodium, silicon, and calcium are dissolved in water when stored in glass.

Gaseous samples may also be kept in glass containers. However, since the container should now be provided with stopcocks, sealing and transport are difficult. Here metal containers are often used. When compressed gases that should be kept under pressure are sampled, metal is obligatory.

Plastic containers, especially polyethylene flasks and jars, are used extensively for transporting samples. The advantages are obvious—they are unbreakable and light—but the disadvantages are the adsorption of trace organics, often irreversible, and the adsorption of trace metals (see, e.g., WE 76, MO 77, HE 77).

Contamination of samples with the plastic, plasticizer, or traces of catalyst is often encountered, especially when organic liquids are stored in plastic bottles.

Gases can be kept in plastic containers only for short periods, since many gases can permeate through thin plastic walls to both sides, so loss of components as well as contamination with oxygen or water are possible.

Sealing of containers against loss or entry of components by using screw caps with seals, rubber stoppers, or Teflon-lined stoppers is usually sufficient. Glass stoppers and cork stoppers should be avoided except when used for dry, coarse solids. Sealing of containers for trade or legal purposes is best done by locking a screw cap with a thread and sealing the knot.

For special purposes, for example, packing samples of highly corrosive or toxic substances and radioactive materials, the appropriate directions should be followed, as given in the literature.

Many samples of food products and samples from the environment are liable to decomposition by bacteria, molds, or enzymes. The most universally applicable treatment is cooling. Samples can be packed in dry ice or stored in a freezer.

Sterilizing by heat is seldom used. Preservation by adding a bactericide (thymol, toluene, mercuric chloride) is often used for environmental samples that are taken in the field or at unattended sampling stations.

A sample is valuable only when it represents the lot it is taken from. This implies that the lot from which it originates must always be known. Because the sample itself as a rule cannot bear the label with its origin (exceptions are samples such as ingots or packed items of food or fine chemicals), the container should bear this information.

All data can be recorded on a label firmly attached to the container, or the data can be held in a protocol, coupled to the container by an identification number.

The sample is identified unambiguously if the time and place are accurately registered. In practice other information may be provided, such as an identification of the lot, for example, the name of a patient, the name of a ship containing the lot, or statement of a traded lot. It is important that the sampler, be it human or machine, can always be identified. In routine work this can be traced if the time and place of sampling are known. In all other situations the sampler should sign the label or the protocol.

Samples for trade or legal purposes must be sealed by the sampler, who should sign the protocol with a witness. In order to circumvent problems when making subsamples for storage or contra-expertise, at least three samples must be provided, separately packed and labeled. Here the sample container often may contain only an identification number, but more data should be recorded in the protocol.

One of the most serious mistakes that can be made (and will be made) in sampling and analysis is the interchange of sample and data. The only infallible method is to distinguish samples by an inherent property, for example, composition. Many ingenious systems are developed to prevent mistakes, especially in clinical chemical laboratories. Most systems trust that identification at the sampling point is correct and stress retention of the identification mark with the sample, for example, by mechanical reading and duplication of labels and samples, or by preventing duplication errors by using error-detecting identification numbers or parity bits.

2.7 QUALITY CONTROL OF SAMPLING

As has been stated in the preceding sections, sample quality can be affected in many ways. The most important sources of sample deterioration are inappropriate sampling schemes (sampling frequency, grab size, or number of grabs), inappropriate sample handling (loss of components, contamination, or decomposition), and inappropriate sample identification (labeling and coupling of sublabels to subsamples).

Forward control can be achieved by proper training of samplers and a thorough setup of sampling schemes. Feedback control is possible by analyzing duplicate samples that are obtained in routine or by random instructions.

Analysis of variance (Section 4.6) may indicate whether the error found is caused by sampling or by analysis.

REFERENCES

AM 56 ASTM Committee E122-56T: "Recommended Practice for Choice of Sample Size to Estimate the Average Quality of a Lot or Process," ASTM, Philadelphia, 1956.

AN 77 O. U. Anders: *Anal. Chem.* 49(1), 33A (1977).

AO 60 AOAC: *Official and Tentative Methods of Analysis*, 9th ed., AOAC, Washington, D.C., 1960.

AR 72 Arbeitskreis: "Automation in der Analyse" *Z. Anal. Chem.* **261**, 1 (1972).

AS 51 ASTM Committee E-11: *ASTM Manual on Quality Control of Materials*, ASTM, Philadelphia, 1951.

AS 54 ASTM Committee E105-54T: "Recommended Practice for Probability Sampling of Materials," ASTM, Philadelphia, 1954.

AS 56 ASTM: *Book of Standards*, ASTM, Philadelphia, 1956.

AS 73 ASTM STP 540: "Sampling, Standards and Homogeneity," ASTM, Philadelphia, 1973.

AT 56 ASTM: *Methods for Chemical Analysis of Metals*, 2nd ed., ASTM, Philadelphia, 1956.

BA 28 B. Baule, A. A. Benedetti-Pichler: *Z. Anal. Chem.* **74**, 442 (1928).

BE 45 H. A. Behre, M. D. Hassialis: *Handbook of Mineral Dressing, Ores and Industrial Minerals*, Wiley, New York, 1945.

BE 54 C. A. Bennet, N. L. Franklin: *Statistical Analysis in Chemistry and the Chemical Industry*, Wiley, New York, 1954.

BO 52 A. H. Bowker, H. P. Goode: *Sampling Inspection by Variables*, McGraw-Hill, New York, 1952.

BU 73 P. C. Buscemi, W. D. Washington: "Sampling, Standards and Homogeneity," ASTM STP 540, ASTM, Philadelphia, 1973, pp. 37–44.

DU 62 A. J. Duncan: *Technometrics* **4**, 319 (1962).

DU 71 A. J. Duncan: *Mat. Res. Stand.* **11**(1), 25 (1971).

EB 75 W. Ebing, G. Hoffmann: *Anal. Chem.* **275**, 11 (1975).

EN 70 J. C. Engels, C. O. Ingamells: *Geochim. Cosmochim. Acta* **34**, 1007 (1970).

EN 73 A. W. England: "Sampling, Standards and Homogeneity," ASTM STP 540, ASTM, Philadelphia, 1973, pp. 3–15.

GI 63 P. M. E. M. van der Grinten: *Control Eng.* **10**(12), 51 (1963).

GR 63 P. M. E. M. van der Grinten: *Control Eng.* **10**(10), 87 (1963).

GR 65 P. M. E. M. van der Grinten: *J. Instr. Soc. Am.* **12**(1), 48 (1965).

GR 66 P. M. E. M. van der Grinten: *J. Instr. Soc. Am.* **13**(2), 58 (1966).

GY 66 P. Gy: *Sampling of Materials in Bulk. Theory and Practice*, Soc. de l'industr. minérale, St. Etienne, France, 1966.

GY 79 M. Gy: *Sampling of Particulate Materials. Theory and Practice*, Elsevier, Amsterdam, 1979.

HA 74 W. E. Harris, B. Kratochvii: *Anal. Chem.* **46**, 313 (1974).

HA 76 G. J. Hahn: *Chemtech.* **1976**, 142 (Feb.).

HE 77 R. W. Heiden, D. A. Aikens: *Anal. Chem.* **49**, 668 (1977).

IN 73 C. O. Ingamells, P. Switzer: *Talanta* **20**, 547 (1973).

IN 76 C. O. Ingamells: *Talanta* **23**, 263 (1976).

JA 76 J.A.O.A.C.: "Report on the International Standard Organization on Fertilizer Sampling and Chemical Analysis," *J.A.O.A.C.* **59**(2), 426 (1976).

KA 66 R. Kaiser: *Z. Anal. Chem.* **222**, 128 (1966).

KA 78 G. Kateman, P. J. W. M. Müskens: *Anal. Chim. Acta* **103**, 11 (1978).

KI 74 W. G. King et al.: *Anal. Chem.* **46**, 771 (1974).

KL 77 R. Klockekämper: *Z. Anal. Chem.* **285**, 345 (1977).

KO 71 D. Kortland, R. L. Zwart: *Chem. Eng.* **1971**, 1, 66 (Nov.).

LE 71 F. A. Leemans: *Anal. Chem.* **43**(11) 36A (1971).

ME 67 A. L. van der Mooren: Thesis, Technical University, Delft, The Netherlands, 1967.

MO 77 J. R. Moody, R. M. Lindstrøm: *Anal. Chem.* **49**, 2264 (1977).

MS 78 P. J. W. M. Müskens, G. Kateman: *Anal. Chim. Acta* **103**, 1 (1978).

MU 78 P. J. W. M. Müskens: *Anal. Chim. Acta* **103**, 445 (1978).

PE 38 F. J. Pettijohn: *Manual of Sedimentary Petrography*, Appleton-Crofts, New York, 1938.

RO 68 D. E. Robertson: *Anal. Chim. Acta* **42**, 533 (1968).

RO 73 R. M. Rosain, et al.: *Anal. Chim. Acta* **65**, 279 (1973).

RY 72 J. C. Ryden, et al.: *Analyst* **97**, 903 (1972).

SK 70 D. A. Skoog and D. M. West: *Fundamentals of Analytical Chemistry*, 2nd ed., Holt, Rinehart and Winston, New York, 1970.

SP 65 H. Sporbeck: *Z. Anal. Chem.* **209**, 60 (1965).

ST 73 A. W. Struempler: *Anal. Chem.* **45**, 2251 (1973).

TO 59 R. C. Tomlinson: *Comprehensive Analytical Chemistry* (C. Wilson and D. W. Wilson, eds.), Vol. 1A, Elsevier, Amsterdam, 1959.

VI 69 J. Visman: *Mat. Res. Stand.* **9**(11), 8 (1969).

VI 71 J. Visman, A. J. Duncan, M. Lerner: *Mat. Res. Stand.* **11**(8), 32 (1971).

WA 59 W. W. Walton, J. I. Hoffman: "Principles and Methods of Sampling," in *Treatise on Analytical Chemistry*, Part I, Vol. 1 Section A (I. M. Kolthoff and P. J. Elving, Eds.), Interscience, New York, 1959, p. 93.

WA 76 R. Wagner: *Z. Anal. Chem.* **282**, 315 (1976).

WE 63 F. J. Welcher (Ed.): *Scott's Standard Methods of Chemical Analysis*, 6th ed., Vol. 2(A), Van Nostrand, New York, 1963, p. 22.

WE 76 H. V. Weiss et al.: *Anal. Chim Acta* **81**, 211 (1976).

WI 64 A. D. Wilson: *Analyst* **89**, 18 (1964).

WO 79 T. C. Woodis, J. H. Holmes, B. Hunter, F. J. Johnson: J.A.O.A.C. **62**, 733 (1979).

ANALYSIS

3.1 CHARACTERISTICS OF AN ANALYSIS

A chemical analysis of a material gives a characterization of that material, or a sample of that material, in terms of chemical composition. A distinction may be made between qualitative and quantitative chemical analysis. A qualitative chemical analysis describes the sample in terms of the identity of the composing elemental parts—atoms or molecules—whereas a quantitative chemical analysis also gives the quantities of each of the composing parts of the sample. Another distinction between analytical analyses can be made based on the type of the characterization: the sample can be described according to the occurrence of types of elements —sorts of atoms—which can be qualitative as well as quantitative. This characterization can be extended with information providing the structure of the sample. Here the link between the various types of atoms is involved, and this type of analysis can be considered as a special kind of qualitative analysis. Apart from chemical analysis, physical analysis is also applied for description of samples. Sometimes it is hard to distinguish between chemical and physical sample analysis, because a chemical composition is often correlated to a physical quantity, for example, in the typification of polymers and other macromolecular compounds. On the other hand, sometimes a chemical analysis is made in order to give a physical description of a sample (DY 71).

3.1.1 Structure of an Analytical Procedure

The analytical procedure nearly always consists of five distinguishable steps:

1. Sampling.
2. Sample preparation.
3. Measuring.
4. Data processing.
5. Testing, controlling, and eventually correcting one or more of the processing stages.

The sampling is described in Chapter 2 together with the considerations involved in a proper sample preparation; the measuring process and the various ways data may be processed are given in Chapter 4. The only thing to be mentioned here is that no analytical procedure is complete without a proper test on each of the stages and of the final result. Based on the type of measurements, analytical processes can be divided into one-dimensional and two-dimensional methods. In spectral analyses where qualitative information on the sample is available beforehand, sometimes the measurement of a single spectral position is sufficient to provide a concentration as well as the complete spectrum. In practice many quantitative analyses are one-dimensional. It is seen that each individual compound or element in a sample requires at least a one-dimensional analysis. Sometimes information on quantitative properties may yield an identification of a sample by a one-dimensional method. For instance, comparison of the measured refractive index of a very pure chemical compound with a file that correlates refractive indexes with molecular structures may provide identification. As a rule the reliability of such an identification is not very high and more one-dimensional methods for other typification should be applied.

Two-dimensional methods of analysis provide a series of measurements as a function of a parameter that characterizes the method. In the continuous wave measuring of a spectrum the energy scale (wavelength, wave number, or magnetic field) is varied in a well-defined and selected manner and at each energy value a measurement (e.g., absorbance, magnetization) is made. The resulting set of data pairs can be represented in a graphic way (Figure 3.1), y (measured quality) as a function of x (selected energy value). Other representations such as dy/dx versus x may also be used such as in EPR spectroscopy, but this does not change the principle of the

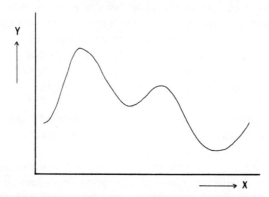

Figure 3.1. Measured quality, y, as a function of energy value, x.

two-dimensional method. A distinction that can be made among two-dimensional methods is qualitative versus quantitative ones. The measurements involved are the same; however, it may not always be possible to extract quantitative information from the spectra (see calibration and standardization procedure). In general quantitative information is found from the y-axis of the graph.

3.2   QUALITY OF AN ANALYTICAL PROCEDURE

In the foregoing section some characteristics of and various distinctions between analytical procedures are made. Given an analytical problem, before a decision can be made on how to select the most appropriate method, more method features must be identified. This discussion aims for a detailed description of the quality of an analytical procedure. In the literature a set of features describing quality aspects of analytical procedures are given (VA 77, GA 78, KA 73) and defined. Some of these properties of features cannot be discussed without also discussing the chemical problem to which the procedure is applied. A method providing a very low detection limit and high accuracy and reproducibility may be adversely influenced because of low selectivity. In practice this may not hamper its application provided the sample preparation step, preceding the measuring step, removes all disturbing elements. Thus the effect of poor selectivity is to require selective sample preparation or application to problems in which other constituents of the sample matrix play no role.

Specific methods with high accuracy and reproducibility may require minimal sample quantities that are relatively high. Here, too, a preprocessing step on sample preparation can be of help, for example, by preconcentrating a sample.

This technique can be applied successfully in some types of gas chromatography where a precondensation column is used to sample the analyte from a large amount of sample of low concentration. The analyte thus obtained is backflushed under other physical or chemical conditions that are applied during the sampling, such as higher column temperatures or evaluation with a gas or liquid stream that has a higher affinity toward the immobile phase than the analyte.

Problems arising from low selectivity or specificity are met when the response of the measuring device is caused by the analyte as well as by one or more components of the sample matrix. In situations where the sample preparation cannot give a proper separation between the analyte and the disturbing components an extended calibration procedure combined with a calculating technique may solve the problem, provided the number of disturbing components is not too high and the signals of the measuring device are well above the noise level (PS 77, PZ 77).

With reference to the cost of an analysis it should be mentioned that, based on salaries of laboratory personnel and prices of modern analytical instruments, a sample preparation of 15–20 minutes by laboratory technicians usually costs considerably more than a measurement of the same period using an advanced instrument (Table 3.6).

These circumstances and the need for quick and reproducible sample analysis, such as is imperative for real-time control of chemical processes with a low time constant, may demand automatic analytical procedures even at high initial instrument costs. In research laboratories, where time and personnel situations are considerably less important than in industrial plants, another analytical procedure may be preferred or selected.

All the remarks of this section should be kept in mind when selecting an analytical procedure, discussed in the next section.

3.2.1 Limit of Detection

Whenever a sample containing a compound in a very low concentration has to be measured by an analytical procedure, the signal from the measuring instrument as a rule will be small. It is difficult to decide whether the signal emerges from the component to be determined or from the inevitable noise produced by the procedure or the instrument. This uncertainty gives rise to the so-called limit of detection. The quality of this limit can be determined by statistical means.

When deciding whether a measured signal originates from the measured property, some incorrect decisions can be made:

1. A decision that the component was present in the sample when in fact it was not: error of the first kind.

2. A decision that the component was not present in the sample when in fact it was: error of the second kind.

Furthermore, it can be argued that there will be a signal that is an indication of the presence of the component, but that does not allow us to make a statement about the true amount of the component. The incorrect decision can now be the following:

- A decision that the amount of component can be given with some precision, when this is not true.
- The opposite decision, that the amount of component cannot be established when in fact this is possible.

In defining the limit of detection a criterion must be selected that is applicable to the decision whether a signal can be used for stating that a component is present/not present and, if present, with what reliability it can be quantified. There are many definitions given in literature. Some, however, using the same criterion, claim different confidence levels. Others define the same, but use different criteria. As a rule the definition of the limit of detection is based on errors of the first kind.

The limit of detection is the smallest observed signal (x) that with a reliability $1-\alpha$ can be considered as being a signal caused by the component to be measured. When the observed signal is smaller than x, however, it cannot be stated that the component is absent. It can only be said with a reliability $1-\beta$ that the concentration of the component will be less than a certain value c_G.

If a blank (concentration of the component to be known, c_0) and a sample (concentration c_1) are analyzed repeatedly, the measuring signals obtained are distributed around $\bar{x}_0$ (blank) or $\bar{x}_1$ (sample) (Figure 3.2).

If the signal with magnitude x_k is used as a criterion for the presence of component c, the probability α that an observed signal $x > x_k$ caused by the blank is

$$\alpha = \int_{x_k}^{\infty} P_0(x)\, dx \tag{3.1}$$

where $P_0(x)$ = probability distribution of x_0
    $P_1(x)$ = probability distribution of x_1
The probability β that a signal will be observed that is less than x_k and

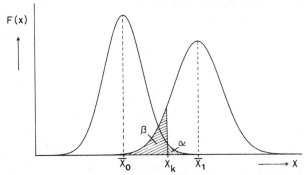

Figure 3.2. Limit of detection. $\bar{x}_0$ = mean signal of blank; $\bar{x}_1$ = mean signal of blank; x_k = criterion value; α = probability that signal of sample is considered as signal of sample; β = probability that signal of sample is considered a signal of blank.

caused by the sample (concentration component c_1) is

$$\beta = \int_0^{x_k} P_1(x)\, dx \tag{3.2}$$

where $\alpha =$ probability of an error of the first kind
$\beta =$ probability of an error of the second kind
If $P_0(x) = P_1(x)$ and $\alpha = \beta$, then x_k is the point of intersection of the two probability distributions.
Now the limit of detection x can be defined as

$$x = \bar{x}_0 + k\sigma_0 \tag{3.3}$$

where $\bar{x}_0 =$ mean of blank signals x_0
$\sigma_0 =$ standard deviation of the probability distribution $P_0(x)$ of x_0
$k =$ factor fixed by $P_0(x)$ and the preferred value of $1 - \alpha$
The composition derived from the signal x is

$$c = \frac{\bar{x}_0 + k\sigma_0}{S} \quad (S = \text{sensitivity}) \text{ (Section 3.2.2)} \tag{3.4}$$

The value of k can be obtained in various ways.

1. When the probability distribution of x_0 is known with sufficient confidence and proved to be Gaussian, α and k are related as shown in Table 3.1. An often-used value of $\alpha = 0.0013$ $(1 - \alpha = 99.87\%)$ results in $k = 3$ so that

$$x = \bar{x}_0 + 3\sigma_0 \tag{3.5}$$

2. When σ is not known but estimated as s from a finite number of observations, Student's-t must be used (t is a correction factor used to compensate for the uncertainty in the estimation of s) (Section 3.2.7).

TABLE 3.1. Values of k

α	k
0.5000	0
0.1587	1.0
0.0228	2.0
0.0062	2.5
0.0026	2.8
0.0013	3.0
0.0005	3.3

When for a certain method of analysis the limit of detection **x** is set equal to the mean of the blank signal,

$$x = \bar{x}_0 \tag{3.6}$$

then $k = 0$ and $\alpha = 0.5$. This implies that a random result is obtained. Now the probability that a signal of a concentration equal to the blank will be considered as a signal of the component is 50%. If the obtained signal $x = \mathbf{x}$ and $k = 3$, we know that

$$x = \mathbf{x} = \bar{x}_0 + 3\sigma \tag{3.7}$$

The probability α that a signal caused by the blank will be considered as a signal caused by the component is 0.0013 (Figure 3.3) (assuming $\sigma_0 = \sigma_1$). β, the probability that a signal caused by the component will be considered to be caused by the blank, is now 50%. If the mean observed signal $x < \mathbf{x}$, we cannot state that $c = Sx$ is smaller than c, the concentration in the blank, because the reliability of this statement is only 0.5. If a signal $x_G = \bar{x}_0 + 6\sigma_0$ (assuming $\sigma_0 = \sigma_1$) and $\mathbf{x} = \bar{x}_0 + 3\sigma_0$, then the situation shown in Figure 3.4 exists.

For a signal $x = \mathbf{x}$ it can be said that the concentration in the sample is less than c_G (if $c_G = Sx_G$) with a probability of 99.87%. This value $c_G = Sx_G$ is sometimes called the "limit of guarantee for purity." c_G here is the lowest concentration that can be detected with a probability of 99.87%. If $\sigma_0 \neq \sigma_1$, x_G is given by

$$x_G = \bar{x}_0 + 3\sigma_0 + 3\sigma_{x_G} = \mathbf{x} + 3\sigma_{x_G} \tag{3.8}$$

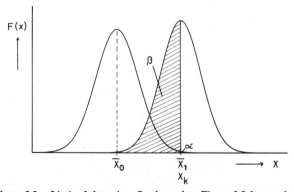

Figure 3.3.   Limit of detection, See legend to Figure 3.2 for symbols.

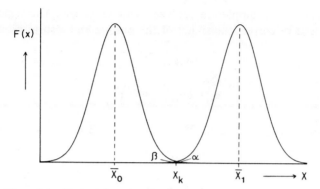

Figure 3.4. Limit of detection. See legend to Figure 3.2 for symbols.

Sometimes a limit of determination is established as

$$x_D = \bar{x}_0 + 10\sigma_x \qquad (3.9)$$

Currie (CU 68) distinguishes three areas (Figure 3.5):

I. The area where detection is not possible at all (tossing a coin would give the same information) or possible only with a limited probability ($\bar{x}_0 < x < x_G$).

II. The area where detection is possible with a sufficient probability ($x > x_G$).

III. The area where quantitative determination is possible ($x > x_D$).

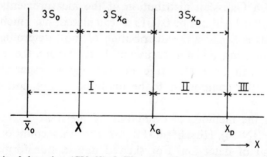

Figure 3.5. Limit of detection (CU 68). I, The area where detection is not possible; II, the area where detection is possible with a sufficient probability; III, the area where quantitative determination is possible. $\bar{x}_0$ = mean of blank signals x_0; x = limit of detection; x_G = limit of guarantee of purity; x_D = limit of determination. Reprinted with permission from *Anal. Chem.* **40**, 588 (1968). Copyright by the American Chemical Society.

Often blank and sample are analyzed simultaneously. The signal of the blank is used to correct the signal of the sample and also to estimate $\bar{x}_0$.

$$x_S = x_{S+x_0} - x_0 \tag{3.10}$$

$$\sigma_{x_S} = \left(\sigma_{x_S+x_0}^2 + \sigma_0^2 \right)^{1/2} \tag{3.11}$$

If $\sigma_{x_S+x_0} = \sigma_0$ then $\sigma_{x_S} = \sqrt{2}\ \sigma_0$. Now $x = \bar{x}_0 + k\sqrt{2}\ \sigma_0$. If more determinations are made (n)

$$x = \bar{x}_0 + \frac{k\sqrt{2}\ \sigma_0}{\sqrt{n}} \tag{3.12}$$

This procedure is applied for enhancing signal-to-noise ratios at the expense of analysis time. In the literature various names are used for the criteria discussed here:

x	Currie:	*a posteriori* upper limit ($k = 1.64$)	(CU 68)
	Kaiser:	Detection limit ($k = 3$)	(KA 65)
x_G	Currie:	*a priori* detection limit ($k = 3.29$)	
	Kaiser:	limit of guarantee for purity ($k = 6$)	
	Liteanu:	Detection limit	(LI 73, LI 75)
	Svoboda:	Nachweisgrenze	(SV 68)
x_D	Currie:	Determination limit ($k = 10.1$)	
	Kaiser:	Determination limit ($k = 10$)	

In the foregoing the values of α and β have been established on the assumption of a Gaussian distribution of the measurements. In practice values of α for $k = 3$ will not be 0.0013 as calculated, but much higher, even up to 0.05. This is because the distribution of most determinations in this low concentration range is non-Gaussian. Furthermore, the inhomogeneity of the sample can be quite large in this region, especially when the distribution of the component to be analyzed is not homogeneous but particulate.

A survey of the methods of estimation of the limit detection has been given by Ingle (IN 74). Hirschfeld (HI 76) gives an account of the practical aspects of limit of detection. For skewed data a transformation can be obligatory (Section 4.1.2, BO 75). In X-ray spectrometry, and other analytical methods based on particle counting, the limit of detection depends more on the counting time than on the background and blank values (PA 72, PL 75).

3.2.2 Sensitivity

A procedure such as an analytical method with one input (the sample) and one or two data outputs can be characterized by the sensitivity S, which is the ratio of the quantitative output and the input, for a given qualitative range. For example,

$$S\,(\lambda = 534\ \text{nm}) = \frac{y}{x} \tag{3.13}$$

where y = extinction at 534 nm
  x = concentration of component x
Often however, y is composed of a part that depends on x and a part independent of x (blank). Furthermore, as a rule y is not linearly proportional to x over the entire range of possible x and y values. Therefore there are good reasons to define

$$S(x, y) = \frac{dy}{dx} \tag{3.14}$$

for the range of x and y values where this relation holds. The range for which S exists and has an unambiguous value is the "dynamic range" of the procedure. Mostly for reasons of convenience, analytical chemists try to develop methods in which S has a constant value in a range as large as possible. This range is called the "linear dynamic range" and is expressed in the orders of magnitude for which S can be considered to be a constant.

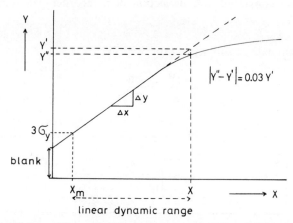

Figure 3.6. Sensitivity and linear dynamic range for output signal y as a function of input signal x.

The dynamic range is limited at the lower level by the value of x where y cannot be distinguished from the noise in y. Here S may have all possible values. The upper limit of the dynamic range will be set by saturation of the detector, $S \to 0$. The linear dynamic range is often defined as the region where $|y' - Sx'| \leqslant 0.03 \, Sx'$, y' being the true signal and Sx' the computed signal, assuming a linear relation between y and x (Figure 3.6).

In situations where the analytical procedure consists of a chain of procedures, the sensitivity S can be subdivided into sensitivities belonging to the comprising operations (Figure 3.7).

$$S_{total} = \frac{dy_n}{dx_n} \cdot \frac{dx_n}{dx_{n-1}} \cdots \frac{dx_2}{dx_1}$$

$$= \frac{dy_n}{dx_n} \cdot \frac{dy_{n-1}}{dx_{n-1}} \cdots \frac{dy_1}{dx_1}$$

$$= S_n S_{n-1} \cdots S_1$$

$$= \prod_{i=1}^{n} S_i \qquad (3.15)$$

Note that the dimension of S depends on the dimension of x and y. If, for example, x is given in grams and y in milliamperes, the sensitivity is expressed in mA/g. Consider the chain of procedures:

- Concentration (g/l) converted into a potential (mV). dim $S = $ mV/(g/l).
- Potential converted into deflection of a recorder pen (cm). dim $S = $ cm/mV.

Then the ultimate sensitivity is expressed in

$$S = \frac{mV}{g} \cdot 1 \cdot \frac{cm}{mV} = \frac{cm \cdot l}{g}$$

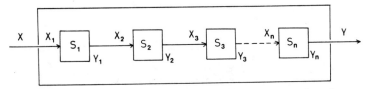

Figure 3.7. Cumulative sensitivity for a signal y resulting from a number of consecutive operations S_i on an input signal x.

The foregoing holds only for static measurements. Under dynamic circumstances S is time-dependent. A sudden disturbance of the value of x is not momentarily followed by an output y; dy/dx is a function of t.

Assuming that y is related to x according to a first-order differential equation, a stepwise disturbance of $0 \rightarrow x$ results in

$$y(t) = Sx\left[1 - \exp\left(\frac{-t}{\tau}\right)\right] \tag{3.16}$$

with τ = time constant of the procedure. This results in a response like that shown in Figure 3.8.

Because the output y is time-dependent, there will be a bias between

$$y(t \rightarrow \infty) = Sx \tag{3.17}$$

and

$$y(t) = Sx\left[1 - \exp\left(\frac{-t}{\tau}\right)\right] \tag{3.18}$$

If this bias should be limited to a fraction Δ of the maximum signal, then

$$y(t \rightarrow \infty) = (1 - \Delta)y \tag{3.19}$$

According to eqs. 3.17 and 3.18

$$(1 - \Delta)Sx = Sx\left[1 - \exp\left(\frac{-t}{\tau}\right)\right] \tag{3.20}$$

$$t = -\tau \ln \Delta \tag{3.21}$$

For $\Delta = 0.01$, t must be at least 4.6τ. However, the measuring instrument has to return to zero for $x \rightarrow 0$, which means that the time lag between two subsequent readings must be at least $t = -2\tau \ln \Delta$. This condition implies that for $\Delta = 0.01$, $t \cong 10$.

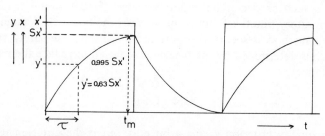

Figure 3.8. Influence of time constant τ on response y of repetitive analysis.

Many instruments of the type that convert a physical quantity into a current or a potential have time constants that range from 100 to 1000 msec. Instruments incorporating handling of mass, such as chromatographs and flow analyzers, have time constants ranging from 100 msec to 10 seconds. Here the time between subsequent analyses can be of the order of 100 seconds. The foregoing remarks apply only to instruments having a first-order response. If time delays are incorporated, as is usual during manipulation of the sample, the time lag between two successive analyses is the sum of this delay time plus the response time of the meter.

3.2.3 Safety

The aspect of safety in analytical chemical practice is very difficult to incorporate quantitatively into the set of criteria. Analytical instruments as a rule are safe, provided they satisfy the safety requirements for electrical instruments. High-energy spectroscopes, emitting X-rays, are liable to give off dangerous radiation, as do instruments fitted with radiation sources, such as gas chromatographs fitted with γ-ray detectors based on Li^3H. If they are properly handled, and the rules set for radioactive materials are satisfied, these instruments do not have a safety risk that has to be considered in deciding whether to apply an analytical method.

The procedures as a rule are safe. Possible exceptions are the destruction of organic material by fusion in sealed vessels or by perchloric acid. If, however, the procedure is carried out according to the directions of the safety standards and the required safety precautions are applied, no abnormal risk is to be expected.

The only risk factor that seems to be important when choosing a method of analysis is the use of dangerous reagents in the preprocessing of the sample. This can be radioactive (in isotope dilution or in the processing of irradiated samples from activation analysis), extremely poisonous, or carcinogenic reagents. All risks can be eliminated to a large extent, either by the use of special equipment such as hoods and glove boxes in special procedures such as decontamination, or specialized training required by "good laboratory practice," but in most cases the cost will be high. When selecting a method it seems appropriate to translate the safety risk into cost, which is determined by the measures required to reduce the risk to an acceptable level.

3.2.4 Cost

Cost as a quality criterion seems a bit odd, but if the criterion used is a "desired property," cost can be a discriminating factor. Cost can be subdivided into several parts and in various ways: overhead and allocated, fixed and variable.

Overhead costs are those of the setup required for analysis. It comprises the laboratory building, joint instruments, service facilities such as glass blower, instrument repair shop, and storage, utilities, and nonallocated personnel such as director, secretary, and cleaning people. Often the overhead costs are not allocated to the laboratory but are fixed costs. Of course somebody has to pay these costs, but as a criterion of the quality of a method of analysis they are not appropriate. Allocated costs are those costs that can and will be paid by the customers of the laboratory. Whether or not these costs are really paid by the customer, they can be used as a criterion to describe the quality of a method of analysis.

Fixed allocated costs are those that are primarily caused by the method of analysis, not directly by the number of analyses: reagents and manpower. Often a trade-off is possible between fixed and variable costs. Mechanization and automation may increase the price of an instrument considerably, but as a rule the manpower needed decreases with increasing degree of automation. The optimum depends on the price of labor, including the unmeasurable benefits of replacing dull work and its increased risk of mistakes. Furthermore, the number of analyses per period and the price of the instrument are important. Another important trade-off is possible in optimizing the cost of accuracy. A rule of thumb is that prices of instruments increase quadratically with a rise in accuracy. Because accuracy can be increased by more frequent analyses, or by more cautious (as a rule, slower) analyses, here again the price of labor and instrument should be incorporated (BE 72, HA 76).

In analysis for control, the most important way of optimizing costs, and thus seeking the most appropriate variables, is to compare costs of analysis for control with the positive or negative costs of the manufacturing process as a result of control. In Section 2.5.4 the influence of accuracy and frequency of analysis on costs was treated.

Data on costs of analysis are scarce in the analytical literature (HO 70). Prices of instruments differ very much in different countries as do labor costs. Moreover, the time required to perform an analysis is seldom given. A rough estimation of the time required between introduction of the sample and release of the analytical result can be obtained by analysis of the method description.

Often the time required for "passive" manipulations such as boiling, waiting, or cooling is described, for instance, "keep boiling for 5 minutes," or "leave on water bath for 5 hours." However, these times are labor-extensive. A rough estimation of the times required is possible, but of little use. "Active" manipulations in routine or semiroutine analysis are divided

TABLE 3.2.   Classification of Analyses

Manipulation	Class	Standard Time (minutes)
Addition, transferring, mixing, Weighing, crushing, dissolving	1	$2\frac{1}{2}$
filtration, heating, cooling, titration	2	5
Extraction, measurement with AA, FES, UV–Vis, AES, XF	3	15

into three categories, the number of manipulations in each class counted and multiplied with a standard time for each class. If some nonallocated time is spread over the manipulations, the values of Table 3.2 are often valid.

Exceptions are NMR (25 minutes) MS, GC, and IR, where the time of analysis depends strongly on the type of measurement.

However, it must be borne in mind that costs of analysis are influenced very much by the organization of the work. Not only overhead, but also "friction losses," waiting time when queues develop, and training and standard of training of personnel, for instance, make possible differences in costs so large that reliable use as quality criterion is very difficult.

Cost as a function of the number of samples taken to establish a fixed precision has been treated by Marcuse (MA 49). The concept of the critical batch size is discussed as a measure useful in deciding whether a laboratory procedure should be automated, or mechanized. Dependence of critical batch size on initial investment, costs of reagents, frequency of sample analysis, lifetime of the equipment, and other important factors to be considered are discussed, using the example of an enzymatic glucose analysis.

Cost is also mentioned in Sections 2.5.3, 2.5.4, 3.1, 3.3.1, and 5.2.1.

3.2.5   Measurability

In studying processes one is interested in an estimation of the quality of equipment and procedures that are applied for measurement and for process control. Criteria that may allow an estimation of this quality are measurability and controllability. In Section 2.5.5 the controllability is given by the numerical reduction of the value of a disturbance of the process σ_x by a controlling action. The remaining value of the disturbance after control, σ, and that before define the controllability factor r according to

$$r^2 = \frac{\sigma_x^2 - \sigma^2}{\sigma_x^2} \qquad 0 \leqslant r \leqslant 1 \qquad (3.22)$$

In order to control a process, that is, minimize the difference between a measured quantity and its desired value $\bar{x}$, one must measure that quantity. The measurability m has been defined in a way resembling the definition equation of the controllability factor:

$$m^2 = \frac{\sigma_x^2 - \sigma_{min}^2}{\sigma_x^2} \qquad 0 \leqslant m \leqslant 1 \qquad (3.23)$$

in which σ_{min}^2 equals the variance of the measured value after application of ideal control. A nonideal control results in an σ-value higher than σ_{min} and thus $r < m$, or, as is stated in Section 2.5.4,

$$r = r_p m \qquad r_p \leqslant 1 \qquad (3.24)$$

The aim of a controlling action is the reduction of disturbances of the process values. The disturbances can be characterized by means of statistical parameters such as

- A distribution-density function that characterizes the amplitudes of periodic process disturbances.
- A correlation function that describes the frequencies of periodic process fluctuations.

The standard deviation of the disturbance σ_x before control and σ after control are related according to

$$\sigma = \sigma_x (1 - r^2)^{1/2} \qquad 0 \leqslant r \leqslant 1 \qquad (3.25)$$

A value $r = 0.5$ results in $\sigma = 0.87\sigma_x$, which implies a very small reduction of the disturbances. A good controlling action produces an r value > 0.9, which means 80% suppression of disturbances.

The frequency of the process fluctuations follows from the autocorrelation function of x_t. According to Section 4.5.2, the autocorrelation gives a relationship between a process value of time t and that of another time. This predicting element can be used for measurability and controllability (MU 78).

When the result of a measurement has been obtained, a prediction can be made of the value of x in the future.

The uncertainty in this prediction is given by the predicted variance

$$\hat{\sigma}_x^2 = \sigma_x^2(1 - \alpha^{2\tau}) \tag{3.26}$$

For first-order, stationary stochastic processes $\alpha = \exp(-1/T_x)$. $\hat{\sigma}_x^2$ equals the predicted variance, and σ_x^2 the variance known from the history of the process.

When the process is sampled regularly with a distance between two consecutive samples T_a, and the time required to obtain the result of the analysis equals T_s, it can be stated that the uncertainty is minimal at the moment the result is produced, a period T_s after the moment of sampling. The uncertainty gradually increases during the period T_a, and sharply decreases when the next result is obtained.

The mean predicted variance over this period is

$$\left(\hat{\sigma}_x^2\right)_m = \frac{\sigma_x^2}{T_a} \int_{T_s}^{(T_a + T_s)} \left[1 - \exp\left(-\frac{2t}{T_x}\right)\right] dt$$

$$= \sigma_x^2 \left[1 - \exp\left(-\frac{2T_s}{T_x}\right)\left(\frac{1 - \exp(-2T_a/T_x)}{2T_a/T_x}\right)\right] \tag{3.27}$$

$$= \sigma_x^2 \left[1 - \exp\left(-\frac{2T_s}{T_x}\right)\right]\exp\left(-\frac{T_a}{T_x}\right)$$

$$m^2 = \frac{\sigma_x^2 - \left(\hat{\sigma}_x^2\right)_m}{\sigma_x^2} = \exp\left(-\frac{2T_s}{T_x}\right)\exp\left(-\frac{T_a}{T_x}\right) \tag{3.28}$$

$$m = \exp\left(-\frac{T_s}{T_x}\right)\exp\left(-\frac{T_a}{2T_x}\right) \tag{3.29}$$

By the same reasoning the expression for m was derived (GI 63, GR 63, GR 65, GR 66) when the accuracy σ_a of the measuring method and the sample size T_{mi} is included:

$$m = \exp\left(-\frac{T_{mi}}{3T_x}\right)\exp\left(-\frac{T_s}{T_x}\right)\exp\left(-\frac{T_a}{2T_x}\right)\left(1 - \frac{\sigma_a T_a^{1/2}}{\sigma_x T_x^{1/2}}\right) \tag{3.30}$$

Figures 3.9 and 3.10 give the m values as a function of σ_a/σ_x and T_a/T_x.

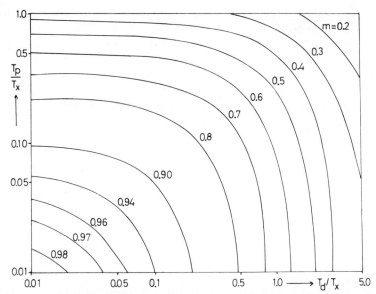

Figure 3.9. Measurability constant m as a function of dead time T_d and measuring time interval T_p for a first-order autoregressive stochastic stationary process with a time constant T_x.

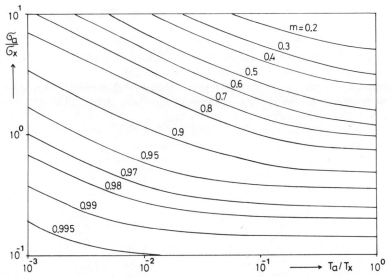

Figure 3.10. Measurability constants m as a function of accuracy of the measuring method σ_a and time lag between two consecutive samples T_a for a first-order autoregressive stochastic stationary process with a time constant T_x and a standard deviation σ_x.

87

These two figures imply a strategy for improvement of m by an attack on the smallest m value.

One must keep in mind that all equations presented here are approximations that are valid for high values of m only, but in situations encountered in practice this is hardly a disadvantage because low m values imply $(1-m^2)^{1/2} \approx 1$ and thus it is immaterial that m may have an approximated value. This is illustrated in the following table, where the reduction of disturbances expressed as σ_{min}/σ_x is tabulated as a function of m according to eq 3.23:

$\sigma_{min}/\sigma_x \, [=(1-m^2)^{1/2}]$	m
0.10	0.995
0.25	0.97
0.50	0.87
0.87	0.50

Examples

The practical implications of these equations can be illustrated with the following example pertaining to the analysis of the nitrogen content of fertilizers. There are many methods available to do this analysis. These methods are compared on the basis of measurability factors (LE 71).

The production of fertilizer in a continuous process sets the following numerical values:

$$T_a = 30 \text{ minutes} \qquad \sigma_x = 1.2\% \text{ nitrogen}$$
$$T_x = 66 \text{ minutes} \qquad m_a = 0.80\%$$

The methods of analysis are listed in Table 3.3 (LE 71). It is seen that the highest m value is given by neutron activation. In a practical situation the cost of the equipment can play an important role in solution of a particular procedure. In static situations the analysis time becomes unimportant. Here the Devarda method and neutron activation give equal information yields.

The influence of averaging the sample obtained during one sampling period is demonstrated as follows: in the process under investigation m_a equals 0.80. Sampling during 30 minutes $(=T_{mi})$ between two consecutive analysis gives $m_{mi} = \exp(-T_{mi}/3T_x) = 0.86$ and $m_a \times m_{mi} = 0.68$. The following list of m values is obtained taking these measurability factors into account:

	m_{total}
Devarda	0.20
Automatic Devarda	0.56
AutoAnalyzer	0.50
Specific electrode	0.50
Roentgen diffraction	0.51
Neutron activation	0.64
γ-Ray absorption	0.59

TABLE 3.3.   Methods for Analysis of Nitrogen in Fertilizer (LE 71)[a]

Criterion and Analytical Technique	Dead time of Analysis (Minutes)	Std. Dev. of Analysis (% N)	$(m_d)_a$	m_n	m
Total N, classical distillation	75	0.17	0.32	0.99	0.24
Total N, DSM automated analyzer	12	0.25	0.84	0.97	0.65
NO_3-N, Technicon AutoAnalyzer	$15\frac{1}{2}$	0.51	0.79	0.92	0.58
NO_3-N, specific electrode	10	0.76	0.86	0.85	0.58
$NH_4NO_3/CaCO_3$ ratio, X-ray diffraction	8	0.8	0.89	0.83	0.59
Total N, fast neutron activation analysis	5	0.17	0.93	0.99	0.74
Specific gravity, γ-ray absorption	1	0.64	0.98	0.88	0.69

Reprinted with permission from *Anal. Chem.* **43**(11), 36A (1971). Copyright by the American Chemical Society.

[a]A sampling frequency of 2 samples/hour is assumed, which means that $m_a = 0.80$; $(m_d)_a$ = measurability caused by analysis time, m_n = measurability caused by accuracy. The related process characteristics are mentioned in the text.

Another example aims for an efficient enhancement of the measurability of a gas chromatographic analysis. It is given that $T_x = 250$ minutes, $T_a = 20.7$ minutes, $T_s = 5.9$ minutes, $\sigma_x = 0.05$, and $\sigma_m = 0.01$. According to eq. 3.30 $m_{total} = m_a * m_x * m_n$ which upon substitution of the appropriate exponentials yields

$$m = \exp\left(-\frac{T_a}{2T_x}\right) * \exp\left(-\frac{T_s}{T_x}\right) * \left[1 - \frac{\sigma_m}{\sigma_x}\left(\frac{T_a}{T_x}\right)^{1/2}\right]$$

$$= 0.88 \tag{3.31}$$

In order to enhance the precision, the peak surface of the component under consideration is normalized to the entire surface of all chromatographic signals, resulting in a σ_m-value about 0. This action enhances T_s to 20.7 minutes because the entire analysis should be finished before the normalization can be calculated. The measurability now becomes $m_a * m_s = 0.88$, which means no improvement at all.

Another procedure that can be applied is backflushing of the column content as soon as the signal of the component under investigation has been found. Now T_a becomes 6 minutes and $m = 0.94$.

Thus it is seen that a reduction of the sampling interval by modification of the method of analysis results in a significant enhancement of the measurability.

3.2.6 Precision

Precision is one of the most, and sometimes exclusively, used criteria for quality of an analytical method. As can be seen from the contents of this book, this is by no means justified. There are many quality criteria that are important in the description of an analytical method or procedure. However, precision has been and will be a very important yardstick to describe an analytical method. Since there are many books on statistics—the field of science that is used to quantify precision—that describe the theory and practical applications, in this book we give just a brief outline (YO 51, AL 69, BE 75, DA 72, DI 69, DU 74, NA 63).

Precision as a quality parameter in analysis can be described qualitatively as the quantity that is a measure for the dispersion of results when an analytical procedure is repeated on one sample. This dispersion of results may be caused by many sources. It is common practice in describing precision only to imply the sources that cause random fluctuations in the procedure. Because the analytical procedure furnishes results that are related to the composition of the sample, the scatter of the results will be around the expected value of the result, if no bias exists. In analytical chemistry this scatter is more often than not of such a nature that it can be described as a normal distribution.

When the results are not normally distributed, a simple transformation often produces a normal distribution (Section 4.1.2). A normal distribution is the frequency distribution of samples from a population that can be described by a Gaussian curve as shown in Figure 3.11. The normal or Gaussian distribution is characterized by the position of the mean, usually denoted as μ, and the half width of the bell-shaped curve at half height, proportional to the standard deviation σ. In practice only a limited number

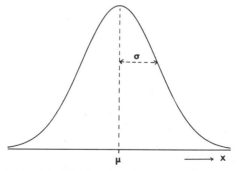

Figure 3.11. The normal probability function characterized by μ = mean and σ = standard deviation.

of observations are made and only estimates of μ and σ are obtained, usually denoted as $\bar{x}$ and s. Often estimates of s can be obtained from the range of the observations. The precision of an analytical method is often defined as the standard deviation s of a series of results obtained from one sample. The probability that a standard deviation s_1 is not significantly different from a standard deviation s_2, obtained from another series of observations on the same sample, should be established by statistical tests. For non-Gaussian distributions of the results, the common tests are not applicable. Therefore the distribution must be tested for normality (Section 4.1.1) and, if it is proved not to be normal, other tests should be applied, or the results must be transformed in such a way that a normal distribution results. In order to distinguish the various sources of scatter, an analysis of variance can be applied (Section 4.6).

The mean is only one type of measure of location of a distribution. Other characteristic quantities are the median and the mode. Since these measures are not often used in analytical quality control, we do not deal with their properties here. The property called simply the "mean" is in fact the arithmetic mean. Other types of mean are the geometric mean and the harmonic mean. The arithmetic mean for n observations is defined as

$$\bar{x} = \frac{1}{n} \sum_{i=1}^{n} x_i \qquad (3.32)$$

For random scatter in the results $\bar{x}$ is also the best available estimate of the population mean μ. As n increases, $\bar{x}$ becomes an increasingly precise estimate of μ. The computation of the mean is straightforward. However, because in the past the calculations often had to be done without calculators or computers, many devices have been used to reduce labor and the chance of errors. Some of the devices that have survived are the following:

1. When the observations consist of several digits, they can often be converted to smaller numbers by subtracting from each a constant quantity. The number of figures that have to be processed is reduced by this method. If the constant subtracted is x_0, then

$$\bar{x} = x_0 + \frac{1}{n} \sum_{i=1}^{n} (x_i - x_0) \qquad (3.33)$$

2. When the observations all end in one or more zeros or are decimal fractions, they may be divided or multiplied by a power of 10, h:

$$\bar{x} = \frac{1}{hn} \sum_{i=1}^{n} (hx_i) \qquad (3.34)$$

3. If a number of observations are available, for each of which the arithmetic mean has already been calculated, the grand mean can be obtained by

$$\bar{x} = \frac{1}{\sum\limits_{i=1}^{k} n_i} \sum_{i=1}^{k} n_i \bar{x}_i \tag{3.35}$$

where the ith group of observations contains n_i observations with group mean $\bar{x}_i$. For equal sample sizes this reduces to

$$\bar{x} = \frac{1}{k} \sum_{i=1}^{k} \bar{x}_i \tag{3.36}$$

The geometric mean is defined as

$$\bar{x}_G = \left(\prod_{i=1}^{n} x_i \right)^{1/n} \tag{3.37}$$

or

$$\log \bar{x}_G = \frac{1}{n} \sum_{i=1}^{n} \log x_i \tag{3.38}$$

that is, the nth root of the product of n observations.

The principal application is in averaging a sequence of ratios. The use of the geometric mean thus amounts to a transformation of the variable to $\log x_i$. As a rule it is better to stick to the transformation and refer to all characteristics as transformed characteristics instead of using a separate name, that is, $\bar{x}_G$.

The harmonic mean is defined by

$$\bar{x}_H = \frac{1}{n} \sum_{i=1}^{n} x_i^{-1} \tag{3.39}$$

This is equivalent to a transformation to the inverse of the variables.

An important descriptor of dispersion is the variance or its square root, the standard deviation. The variance of a population is the mean squared deviation of the individual values from the population mean and is denoted by σ^2. When the variance is estimated from a finite set of data, the symbols var or s^2 are used. The standard deviation s, the positive square root of the variance, has the same dimension as the variance and therefore as a rule is used in the final result.

In computations, however, it is often more relevant to use the variance:

$$\sigma^2 = \frac{1}{n} \sum_{i=1}^{n} (x_i - \mu)^2 \qquad (n \to \infty) \tag{3.40}$$

The estimate s^2 of σ^2 is

$$s^2 = \frac{1}{n-1} \sum_{i=1}^{n} (x_i - \bar{x})^2 \tag{3.41}$$

An often-used relationship that can be derived from eq. 3.41 is

$$s^2 = \frac{\sum_{i=1}^{n} x_i^2 - (1/n)\left(\sum_{i=1}^{n} x_i\right)^2}{n-1} \tag{3.42}$$

The denominator $n-1$ represents the number of independent observations when the grand total of the results, $\sum_{i=1}^{n} x_i$ is used, for one of the observations can be calculated when all other results and $\sum_{i=1}^{n} x_i$ are known. The term $n-1$ is called the number of degrees of freedom.

A measure of dispersion that has no theoretical value but is sometimes used as a quality criterion is the coefficient of variation, the standard deviation expressed as a percentage of the arithmetic mean. No symbol for this criterion is in common use.

$$\text{Coefficient of variation} = \frac{100s}{\bar{x}} \tag{3.43}$$

3.2.7 Accuracy

The concept of accuracy is one of the most difficult topics in analytical chemistry. Not only is the definition vague and difficult to interpret, but also the theoretical background of the methods to compute accuracy and the methods to estimate accuracy are complicated, ambiguous, and not generally accepted.

Some of the definitions encountered in analytical chemical literature are given here. In a paper in *Analytical Chemistry* on statistical terms it is stated (AN 75):

> Accuracy normally refers to the difference (error or bias) between the mean, $\bar{x}$ of the set of results and the value $\hat{x}$, which is accepted as the true or correct

value for the quantity measured. It is also used as the difference between an individual value x_i and $\hat{x}$. The absolute accuracy of the mean is given by $\bar{x} - \hat{x}$ and of an individual value by $x_i - \hat{x}$.

The commission on analytical nomenclature of IUPAC (FE 69) used the word "bias" to denote accuracy:

Bias, the mean of the differences of the results from the true value. This equals the difference between the mean of a series of results and the true value.

The commission on spectrochemical and other optical procedures for analysis of IUPAC (AN 76):

Accuracy relates the agreement between the measured concentration and the "true value." The principal limitations on accuracy are: a. random errors, b. systematic errors due to bias; bias represents the positive or negative deviation of the mean analytical result from the known or assumed true value and c. (in multicomponent systems) the treatment of interelement effects may involve some degree of approximation.

Cali (CA 75) stated:

An accurate measurement is one that is both free of systematic error and precise. Three major components of the measurement process must be attended to:
• an agreed-on system of measurement
• methods of demonstrated accuracy
• reference materials.

In our opinion "accuracy" cannot define a quantity; it only refers to the degree of attainability of the theoretical concept of "accurate." An accurate measurement is one that is both free of bias and precise. Bias is the mean of the differences of the results from the known or assumed true value. A precise measurement is one that shows no scatter in the results when repeated.

Although accuracy cannot be quantified it is possible to measure some properties that are related to the concept of accuracy. Precision can be measured, although a description of the circumstances of the measurement is required (Section 3.2.6). Bias can be estimated in some ways. In analytical chemistry often the only requirement for a result is that it be comparable with other results. It is an exception when the true value must be known. Therefore, as a rule results of analysis are compared with the results obtained from the analysis of "standards" or "reference materials,"

materials with known or assumed properties. The property of the standard to be known, $\hat{x}$, can be obtained in various ways.

1. *Primary Standards.* The theoretical composition of material of high purity can be used. The purity must be ascertained by independent methods, for example, by melting or boiling points, X-ray diffraction, and electrical properties. These types of standards are used in spectroscopy and titrimetry.

2. *Secondary Standards.* The composition of these materials is measured by agreed-on methods by institutes equal to this task. A well-known institute in the United States that submits certified material is the National Bureau of Standards (NBS) (CA 75). Where possible, the NBS certifies the numerical value of the property(ies) under investigation as "accurate"; that is, within stated uncertainty they are "true values." This is accomplished by three alternative methods:

a. Measurement by a method of known and demonstrated accuracy performed by two or more analysts working independently.

b. Measurement by two or more independent and reliable methods whose estimated inaccuracies are small, relative to accuracy required for certification.

c. Measurement via a qualified network of laboratories.

3. *Standard methods.* The composition of the material to be known can be obtained by applying an agreed-on method of analysis, for instance, a method issued by the International Standardization Organization (ISO) and the American Society for Testing Materials (ASTM). The value obtained by analysis according to a standard method can be assumed to be a "true value." This method of assuring accuracy is often used in trade.

4. *Mean is True.* The mean of the results of a number of independent, selected laboratories can be assumed to be the "true value." It must be ascertained that all participants use comparable methods of data presentation and data handling (YO 75).

When the degree of accuracy is estimated two related questions are often encountered:

1. When is the bias significant, or what is the probability that a found bias is real?

2. How can bias be distinguished from random errors?

The latter problem is dealt with in the sections on analysis of variance (Section 4.6) and sequential analysis (Section 3.5). It should be stressed that random errors may seem to be bias, in particular when the random error is large compared to the bias. If a sample that is too small is taken from an inhomogeneous object, for example, the result of the analysis may deviate appreciably from the results of a thoroughly mixed standard. Repeating the analysis with a standard method may indicate that no real bias is present. Actually here the limited number of samples has caused the apparently significant bias. Even when the analytical method shows no bias at all, the method cannot be denoted accurate when the precision is insufficient. This also implies that it is useless to look for an accurate method of analysis in situations where sampling shows a bias or random error that is large compared to bias and precision of the method of analysis.

The so-called Student's-t test can be applied to test the significance of a detected difference between the measurement results of sample and standard.

This method is based on the assumption that the mean value of a series of observations shows a distribution around the mean value estimated from an infinitely large series, just as individual observations show a distribution around their mean. The distribution of the means has a probability density function that resembles the well-known Gaussian distribution for individual observations. This distribution is called the t distribution. Values of t are tabulated (see Table 3.4) for the probability that a particular mean belongs to a given population of means. If we state

H_0: $\bar{x} = \mu_0$ with $\bar{x} =$ mean of tested population

H_1: $\bar{x} \neq \mu_0$ with $\mu_0 =$ true mean of a given total population

then, if H_0 is true,

$$t\left(\frac{\alpha}{2}, n-1\right) \leqslant \frac{\bar{x} - \mu_0}{s/\sqrt{n}} \leqslant t\left(\frac{1-\alpha}{2}, n-1\right) \tag{3.44}$$

with $n - 1 =$ degrees of freedom
     $s =$ standard deviation
     $\alpha =$ significance level

or

$$t\left(\frac{\alpha}{2}, n-1\right)\sqrt{s^2/n} \leqslant \bar{x} - \mu_0 \leqslant t\left(\frac{1-\alpha}{2}, n-1\right)\sqrt{s^2/n} \tag{3.45}$$

This equation can be applied only when μ_0 is known. If μ_0 has to be estimated with a finite number of observations (e.g., $\bar{x}_2$ from n_2

TABLE 3.4 Probability Points of the t-Distribution (Single-Sided)

ϕ	\multicolumn{5}{c}{P}				
	0·1	0·05	0·025	0·01	0·005
1	3·08	6·31	12·70	31·80	63·70
2	1·89	2·92	4·30	6·96	9·92
3	1·64	2·35	3·18	4·54	5·84
4	1·53	2·13	2·78	3·75	4·60
5	1·48	2·01	2·57	3·36	4·03
6	1·44	1·94	2·45	3·14	3·71
7	1·42	1·89	2·36	3·00	3·50
8	1·40	1·86	2·31	2·90	3·36
9	1·38	1·83	2·26	2·82	3·25
10	1·37	1·81	2·23	2·76	3·17
11	1·36	1·80	2·20	2·72	3·11
12	1·36	1·78	2·18	2·68	3·05
13	1·35	1·77	2·16	2·65	3·01
14	1·34	1·76	2·14	2·62	2·98
15	1·34	1·75	2·13	2·60	2·95
16	1·34	1·75	2·12	2·58	2·92
17	1·33	1·74	2·11	2·57	2·90
18	1·33	1·73	2·10	2·55	2·88
19	1·33	1·73	2·09	2·54	2·86
20	1·32	1·72	2·09	2·53	2·85
21	1·32	1·72	2·08	2·52	2·83
22	1·32	1·72	2·07	2·51	2·82
23	1·32	1·71	2·07	2·50	2·81
24	1·32	1·71	2·06	2·49	2·80
25	1·32	1·71	2·06	2·48	2·79
26	1·32	1·71	2·06	2·48	2·78
27	1·31	1·70	2·05	2·47	2·77
28	1·31	1·70	2·05	2·47	2·76
29	1·31	1·70	2·05	2·46	2·76
30	1·31	1·70	2·04	2·46	2·75
40	1·30	1·68	2·02	2·42	2·70
60	1·30	1·67	2·00	2·39	2·66
120	1·29	1·66	1·98	2·36	2·62
∞	1·28	1·64	1·96	2·33	2·58

observations), then

$$t\left(\frac{\alpha}{2}, n_1+n_2-2\right)\left[\left(\frac{1}{n_1}+\frac{1}{n_2}\right)s^2\right]^{1/2}$$

$$\leqslant \bar{x}_1 - \bar{x}_2 \leqslant t\left(\frac{1-\alpha}{2}, n_1+n_2-2\right)\left[\left(\frac{1}{n_1}+\frac{1}{n_2}\right)s^2\right]^{1/2} \quad (3.46)$$

As is obvious, the proof that $\bar{x}_1$ and $\bar{x}_2$ differ with a certain significance, does not prove that the method with $\bar{x}_1$ is not accurate, even if $\bar{x}_2$ has been estimated from a large number of observations. $\bar{x}_2$ is just the mean of observations. Only if it has been ascertained by many means that $\bar{x}_2$ is the true mean, or μ_0 does disagreement between $\bar{x}_1$ and $\bar{x}_2$ tell us something about accuracy.

3.2.8 Measuring Functions and Calibration Functions

During the analysis of a sample, a number of actions are performed, resulting in one or more measured data that are used to calculate quantitative analytical results. From this point of view analysis may be considered to be a process with input parameters on composition, denoted as concentrations x_i, and with output parameters measured data y_i:

$$\xrightarrow[i=1,2,\ldots,n]{x_i} \boxed{\begin{array}{c} \text{PROCEDURE} \\ \text{(Analysis)} \end{array}} \xrightarrow[i=1,2,\ldots,n]{y_i}$$

In order to evaluate from y_j the value for x_i, the functional relationship between y_j and x_i should be known. The establishment of this functional relationship is called calibration. An analytical method for analysis of n compounds $x_1, x_2, \ldots, x_n$ of a sample should produce $y_1, y_2, \ldots, y_m$ measured data $(m \geqslant n)$.

These data should be mutually independent, as are the concentrations of the n compounds. The functional relationship between measured data y and composition x is called calibration function. In order to apply a given analytical procedure for chemical analysis, each measurement y_i should correspond to a composition x_i according to

$$x_i = y_i \qquad (y_j, \text{ all } x \text{ except } x_i) \tag{3.47}$$

In general this condition implies that in order to measure one concentration value x_1, all other n concentrations x_j $(j \neq i)$ should be known, or that these values should be measured by collecting at least $n-1$ more y_i-values. This implies for quantitative analysis the solution of m equations with n unknowns $(m \geqslant n)$.

$$
\begin{aligned}
x_1 &= g_1 \quad (y_1, y_2, \ldots, y_m) \\
x_2 &= g_2 \quad (y_1, y_2, \ldots, y_m) \\
&\;\;\vdots \\
x_n &= g_n \quad (y_1, y_2, \ldots, y_m)
\end{aligned}
\tag{3.48}
$$

These functions are named the analysis or measuring functions G; the calibration functions are $y = F * x$. F contains no y-terms and G contains no x-terms. In general, the g-functions are summations of y-values, multiplied with weight functions u, and the f-functions are summations of x-values, multiplied by weight functions, say, v. In linear situations, which are the only ones that can be handled in practice, at most $m * n$ weighting coefficients should be known. Each of the components n should be used for calibrating, thus producing m y-values per component.

In order to solve for n unknowns only n independent equations are required, which means that out of the m measurements, $m - n$ can be eliminated. The number of calibration samples thus remains n, but the number of calibration measurements may diminish to $n - y$ values. In nonlinear situations in general there will be no solution.

This leaves the problem of what rows and what columns should be removed out of G and F. As a rule of thumb, one may state that the resulting $n * n$ matrix should be such that its determinant has a maximal value. This means that the determinant has highest possible elements on the diagonal and smallest possible elements on the nondiagonal positions. A generally accepted method to reduce the number of columns is to multiply the matrices with their transpose (rows and columns interchanged).

This method is illustrated in the following example: in a spectrophotometric measurement on chlorine and bromine dissolved in a suitable solvent, extinction coefficients a_x are known and the extinctions obtained are shown in Table 3.5. In a mixture of Cl_2 and Br_2 one measures with an optical path length d (cm).

$$y_\lambda = \left(\log \frac{I_0}{I} \right)_\lambda = \left[C_{Cl_2} * a_{Cl_2}(\lambda) + C_{Br_2} * a_{Br_2}(\lambda) \right] * d \qquad (3.49)$$

In order to reduce the six equations with two unknowns (the concentrations C_{Cl_2} and C_{Br_2}) we multiply the left-hand side as well as the right-hand side with the

TABLE 3.5   Extinction Coefficients of Chlorine and Bromine

λ (nm)	a_{Cl_2} (1/mol·cm)	a_{Br_2} (1/mol·cm)	$\log\left(\dfrac{I_0}{I}\right)_\lambda = y_\lambda$ Mixture (optical path length 1 cm)
455	4.5	168.	0.1995
417	8.4	211.	0.2698
385	20.	158.	0.2980
357	56.	30.	0.4220
333	100.	4.7	0.7047
312	71.	5.3	0.5023

transpose $a^\dagger$ of the extinction matrix a (see also Section 4.3.2).

$$a^\dagger * y_\lambda = a^\dagger * a * x * d \tag{3.50}$$

which becomes

$$
\begin{vmatrix} 4.5 & 8.4 & 20. & 56. & 100. & 71. \\ 168. & 211. & 158. & 30. & 4.7 & 5.3 \end{vmatrix} * \begin{vmatrix} 0.1995 \\ 0.2698 \\ 0.2980 \\ 0.4220 \\ 0.7047 \\ 0.5023 \end{vmatrix} =
$$

$$
\begin{vmatrix} 4.5 & 8.4 & 20. & 56. & 100. & 71. \\ 168. & 211. & 158. & 30. & 4.7 & 5.3 \end{vmatrix} * \begin{vmatrix} 4.5 & 168. \\ 8.4 & 211. \\ 20. & 158. \\ 56. & 30. \\ 100. & 4.7 \\ 71. & 5.3 \end{vmatrix} * \begin{vmatrix} C_{Cl_2} \\ C_{Br_2} \end{vmatrix} \tag{3.51}
$$

or

$$
\begin{vmatrix} 138.889370 \\ 156.162080 \end{vmatrix} = \begin{vmatrix} 18667.81 & 8214.7 \\ 8214.7 & 98659.18 \end{vmatrix} * \begin{vmatrix} C_{Cl_2} \\ C_{Br_2} \end{vmatrix} \tag{3.52}
$$

It is seen that the determinant F is well conditioned because $8214.7 < 18667.81$ and < 98659.18.

Application of Cramer's rule (provided $\det \neq 0$) gives:

$$C_{Cl_2} = \frac{F^{1,1}}{\|F\|} * x = \frac{F_{2,2} * x_1 - F_{1,2} * x_2}{F_{1,1} * F_{2,2} - F_{1,2} * F_{2,1}} = 0.0070 \ (\text{mol } Cl_2/l) \tag{3.53}$$

and

$$C_{Br_2} = \frac{F^{2,1}}{\|F\|} * x = \frac{-F_{1,2} * x_1 + F_{1,1} * x_2}{F_{1,1} * F_{2,2} - F_{1,2} * F_{2,1}} = 0.010 \ (\text{mol } Br_2/l) \tag{3.54}$$

in which $F = a^\dagger * a * d$.

3.2.9   Selectivity and Specificity

In quantitative analysis the measurement of one particular component of a mixture may be disturbed by other components of the mixture. This means that the measurement is nonspecific for the compound under investigation.

With a completely specific method for a compound the concentration of that component can be measured, regardless of what other compounds are present in the sample. The other components do not produce an analytical signal. A method of analysis is called completely selective when it produces correct analytical results for various components of a mixture without any mutual interaction of the components. A selective method thus is composed of a series specific measurements. In situations where the linear system holds,

$$y = F * x \tag{3.55}$$

the measurement for component k is specific when only the terms $f_{i,k}$ ($i = 1, \ldots, n$) are nonzero. Here the matrix is called F_{spec} and the value of y is a function of x_k only.

For a completely selective method each row of matrix F contains only one nonzero element. By interchanging rows one may get a matrix F with all nonzero elements on the diagonal. This matrix is called F_{sel}. In practical situations F_{spec} and also F_{sel} contain other nonzero elements. These elements, however, have considerably smaller numerical values than the corresponding $f_{i,k}$ values. The undesired nonzero elements of F_{sel} and F_{spec} can be used for a quantitative description of selectivity and specificity (KA 73). Kaiser proposes a method for calculation of selectivity based on the concept that a procedure becomes more selective with a relative decrease of the nondiagonal elements with respect to the diagonal ones.

In contrast, we prefer the definition of a selectivity number ζ_i, for each row i of F_{sel} by

$$\zeta_i = \frac{|f_{i,i}|}{\sum\limits_{j<i}^{n} |f_{i,j}|} \tag{3.56}$$

It is seen that selectivity becomes maximal when ζ_i approaches infinity. We also prefer specificity for compound k, denoted by γ_k, defined as

$$\gamma_k = \frac{|f_{kk}|}{\sum\limits_{j \neq k}^{n} |f_{j,j}|} \tag{3.57}$$

where a high value of γ_k means a high specificity of the analytical method for compound k.

In our example

$$\zeta_{Cl_2} = \frac{18667.81}{8214.7} = 2.27$$

$$\zeta_{Br_2} = \frac{98659.18}{8214.7} = 12.05$$

$$\gamma_{Cl_2} = \frac{18667.81}{98659.18} = 0.19$$

and

$$\gamma_{Br_2} = \frac{98659.18}{18667.81} = 5.29$$

which means that both the selectivity and specificity of this method are better for bromine than for chlorine.

3.3   SELECTION OF AN ANALYTICAL PROCEDURE

One of the difficulties that confronts an analytical chemist when a particular problem has to be solved is how to select a procedure out of the growing number of analytical methods available. During recent years the number of methods and their applications has increased to such an extent that it seems impossible to familiarize oneself with all possibilities by one's own experience or by study. In common practice this situation results in applying exclusively those methods and procedures that are familiar and well known. Apart from this, one should be aware of the simplification of one's choice because of the limitations of the instrumentation of a particular laboratory. Another aspect of the problem is that during a chemistry course it is not possible for a student to receive a complete and lucid picture of all existing analytical techniques and their application. These problems are well realized among analytical chemists, as may be concluded from a quotation of Kaiser (KA 70):

> At present there is no systematic approach to answer the question—which one may be the best analytical procedure to solve an analytical problem. We are relying on the knowledge and experience of the analyst, we are scanning the literature, we are dependent on habits and fashions. Isn't that muddling through?

Therefore one should ask oneself, "Is it possible to simplify the selection of a method for solving a particular problem according to objective rules?" In order to formulate a set of objective rules one may start by listing criteria for analytical procedures such as limit of detection, selectivity, and accuracy. These criteria can be used for mapping all available procedures in a multidimensional space, where the number of criteria equals the number of dimensions of that space. By varying a procedure a particular analytical technique such as atomic absorption may cover a range of a certain criterion such as detection limit. In the multidimensional methods representation space this results in a multidimensional plane or volume enclosing an area in which that technique can be successfully applied.

Sometimes different techniques are represented by areas that partly overlap, which means that in the overlapping area both techniques can be applied.

The same kind of representation may be developed for a problem space, in which each analytical problem, described by a set of problem criteria such as chemical compound analyzed for, required accuracy, and expected grade, is positioned. By mapping the methods space on the problem space one may see whether the problem is positioned within the boundaries of the method and conclude what method or methods are applicable. Tests should be made as to whether methods selected according to this procedure prove to be better than methods selected randomly.

3.3.1 Cost of Analysis

A selection criterion that requires special attention is the cost of a particular analysis. In order to estimate the cost C of an analysis one may use the following equation:

$$C = \tau_a f + 1.74 * 10^{-6} * F * \tau_n \qquad (3.58)$$

where the first term on the right-hand side denotes the operating cost per analysis based on labor and laboratory overhead costs expressed in money per time unit f, and τ_a, the handling and preprocessing time by laboratory personnel, is expressed in the same time units (e.g., minutes per analysis). The second term denotes the capital investment in analytical equipment F, reduced by the number of analyses performed during its economic life with a "normal" work load. This amount of money is multiplied by τ_n, the duration of one analysis, expressed in the proper time units (e.g., minutes). For on-line analytical instruments, the number of analyses performed in its economic life is fixed and may be calculated from the frequency with which samples are analyzed, for example, for process control. The number of analyses performed with a laboratory instrument is not quite fixed because it depends on the number of analyses required. Here F should be estimated. It may be interesting to see how total cost C is composed for a

TABLE 3.6.   Prices of Instruments

Instrument Class	Investment (arbitrary units)									
	AA	UV/Vis	Tit	RöFlu	AES	GLC	FES	MS	IR	PMR
1	5	2	0	—	—	10	5	50	15	—
2	10	10	2.5	100	30	15	10	100	25	—
3	20	20	5	150	60	20	20	200	35	100
4	30	40	7.5	200	120	30	30	350	60	200
5	50	80	10	300	—	50	50	500	100	300

value of f of 80 money units/hour and an investment of 10^5 money units when τ_a equals 30 minutes and τ_n 15 minutes:

$$C = 40 + 2.03 = 42.03 \tag{3.59}$$

which shows labor to be much more expensive than investments.

The cost of investment is the capital investment in equipment. This amount of money mainly depends on the quality of the instrument under consideration. As a yardstick Table 3.6 is used, based on five classes of instruments, ranging from primitive (1) through "normal" to advanced (5).

3.4   QUALITY CONTROL

In analytical work that has to be controlled, two levels can be distinguished. The first level pertains to the analytical procedure. Most of the quality parameters described earlier in this book are on the procedure level, for example, sampling size, sampling frequency, limit of detection, sensitivity, and precision. The second level belongs to the laboratory. When an analytical procedure is performed, not only are the results of various workers and instruments compared, but also the results of other laboratories where comparable procedures are applied.

In this chapter some methods of controlling the quality, based on the axioms mentioned, are described.

3.4.1   Ruggedness Test

This test, described in full by Youden (YO 72), discloses the influence on the quality of the analytical method of small alterations in the procedure. By ranking these influences it can be demonstrated how susceptible the method is to minor fluctuations and which parts of the procedure should be given extra attention. Such things as the source and age of reagents, concentration of reagents, rate of heating, thermometer errors, humidity, and many other factors may be involved. One laboratory technician may make up a supply of nominally 1 kmol/m³ acid and in fact achieve a concentration of 0.95. Another's solution may be 1.03 kmol/m³.

The only way to attack errors originating from such sources, which are difficult to discover, is to introduce deliberately minor reasonable variations in the method and observe what happens. On first sight this appears to throw extra work onto the analyst, but if the program is carefully planned, a small amount of work suffices. When the appropriate working scheme is used, eight measurements suffice to screen seven variables. The basic idea is not to study one alteration at a time but to introduce several changes at once, in such a manner that the effects of individual changes can be ascertained.

Let A, B, C, D, E, F, and G denote the nominal values for seven parameters that may affect the result of the procedure if their nominal values are slightly changed. Let their alternative values be denoted by the corresponding lowercase letters a–g. Now the conditions for performing a determination will be completely specified by a combination of these seven characters, each character being either a capital or lowercase. There exist 2^7 or 128 different combinations. Fortunately, it is possible to select a subset of eight of these combinations that have a fine balance between capital and lowercase letters. The particular set of combinations is shown in Table 3.7.

Suppose now the quality parameter to be measured is the accuracy. Now analytical results following from each of the combinations are designed by

TABLE 3.7. **Eight Combinations of Seven Factors Used to Test the Ruggedness of an Analytical Method**

Factor	Combinations							
	1	2	3	4	5	6	7	8
A/a	A	A	A	A	a	a	a	a
B/b	B	B	b	b	B	B	b	b
C/c	C	c	C	c	C	c	C	c
D/d	D	D	d	d	d	d	D	D
E/e	E	e	E	e	e	E	e	E
F/f	F	f	f	F	F	f	f	F
G/g	G	g	g	G	g	G	G	g
Result	s	t	u	v	w	x	y	z

Reprinted with permission from W. J. Youden and W. H. Steiner, *Statistical Manual of the AOAC*, Association of Official Analytical Chemists, Washington, D.C., 1975.

the letters s–z. Observe that determinations 1–4 were run with factor A at level A and determinations 5–8 with the factor at level a. Observe that in the combinations with A's the other six factors occur twice at the capital level and twice at the lowercase level. This is also true for the combinations with a. The effects of these factors on the analytical procedure, if present, consequently level out. After the results of all combinations are averaged, a comparison of the average $(s+t+u+v)/4$ with the average $(w+x+y+z)/4$ reveals the effect of changing A to a.

A problem may arise when more or less than seven factors must be examined. Schedules are available for 11 and 15 factors (PL 46). When other sets of factors should be studied an excellent method is described by Youden. Suppose only six factors are to be explored. By that event, associate with the seventh factor g some meaningless operation such as solemnly picking up an object, for example, a beaker, looking at it intently, but not impolite, and setting it down again cautiously. Omit this meaningless operation for the determinations that involve factor G. (Be sure to look at the average difference between the G's and g's, because if it is large an explanation should be sought!)

The differences between the two alternatives of each factor can be estimated with Table 3.8. List the differences V_A–V_G in order of magnitude. If one or two factors show an effect, the differences produced on a variation of these factors will be substantially larger than the group of differences associated with the other factors. This ranking is a direct guide to the analytical method's sensitivity to modest alterations in the factors. When no dominant differences are found, the most realistic measure of the analytical error S' is calculated by

$$S' = \left[\tfrac{2}{7}\left(V_A^2 + V_B^2 + V_C^2 + V_D^2 + V_E^2 + V_F^2 + V_G^2\right) \right]^{1/2} \tag{3.60}$$

This calculation can be tested against the standard deviation s obtained in

TABLE 3.8.   **Ranking Scheme for the Ruggedness Test**

$$V_A = \tfrac{1}{4}(s+t+u+v) - \tfrac{1}{4}(w+x+y+z) \equiv (A-a)$$
$$V_B = \tfrac{1}{4}(s+t+w+x) - \tfrac{1}{4}(u+v+y+z) \equiv (B-b)$$
$$V_C = \tfrac{1}{4}(s+u+w+y) - \tfrac{1}{4}(t+v+x+z) \equiv (C-c)$$
$$V_D = \tfrac{1}{4}(s+t+y+z) - \tfrac{1}{4}(u+v+w+x) \equiv (D-d)$$
$$V_E = \tfrac{1}{4}(s+u+x+z) - \tfrac{1}{4}(t+v+w+y) \equiv (E-e)$$
$$V_F = \tfrac{1}{4}(s+v+w+z) - \tfrac{1}{4}(t+u+x+y) \equiv (F-f)$$
$$V_G = \tfrac{1}{4}(s+v+x+y) - \tfrac{1}{4}(t+u+w+z) \equiv (G-g)$$

the usual way:

$$\bar{x} = \frac{s+t+u+v+w+x+y+z}{8} \tag{3.61}$$

$$s = \left(\frac{(s-\bar{x})^2 + (t-\bar{x})^2 + \cdots + (z-\bar{x})^2}{7} \right)^{1/2} \tag{3.62}$$

If the standard deviation thus found is unsatisfactorily large, it seems appropriate to pay more attention to the factors that are high ranking.

The method described for the quality parameter accuracy can be used equally well for other quality parameters, though some of them are difficult to estimate.

3.4.2 Control Charts

One of the oldest methods used to monitor and control quality of a procedure, in particular, the accuracy and precision, is the control chart (DA 72). A control chart is a graphic representation on which the values of the quality characteristic under investigation are plotted sequentially. The type of chart as devised by Shewhart (SH 31) consists of a central line and two limit lines spaced above and below the central line. These are usually termed the inner and outer control limits. The distribution of the plotted values with respect to the control limits provides valuable statistical information on the quality characteristic being studied. Another type of chart shows the cumulative differences of the quality characteristic from a target level also plotted sequentially. This chart is known as a cumulative sum chart. Of course a control chart is not an end in itself; it is an aid in prediction and a basis for action. The control chart may take a variety of forms. This chapter is concerned with the basic principles; detailed explanations and examples can be found in books on quality control of production processes (DA 72).

Shewhart Control Charts

If the chart is used for subgroup averages, the main steps to be taken are as follows. Use the individual readings of the property on which the quality control is based, for example, the composition of a standard sample analyzed according to the analytical method to be controlled. Plot the readings in the order in which they were obtained. Sometimes the observations can be subdivided into a number of rational subgroups, within which there is reason to believe that only chance causes of variation have

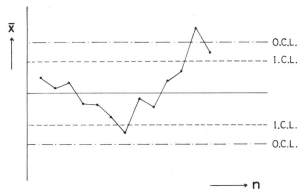

Figure 3.12. Control chart.

operated but between which assignable causes may have operated. Thus a subgroup may consist of the results emanating from one repeated experiment if the results are obtained with a short time span and by one person. If none of the plotted points falls outside the outer control limits it is concluded that the data are statistically uniform; that is, no major assignable causes of variation have intervened during the period (Figure 3.12).

The drawing of the limiting lines for subgroup averages is based on the following principles. If individual observations from the same population, for example, results obtained from a standard sample, are distributed about a mean level μ with a standard deviation σ, then the averages of subgroups, each containing n individuals, are distributed around μ with a standard error of $\sigma/\sqrt{n}$. Whatever the distribution of the individuals, the distribution of subgroup averages approaches normality as n increases. Even when n is as small as four or five, the distribution of averages is close to normal provided the distribution of the individuals is not extremely asymmetric. When subgroups are composed of readings obtained in experiments repeated within a short time span, it is not probable that the distribution is asymmetric. Provided the data are statistically uniform, only one subgroup mean in 40 will on the average lie above the limit $\mu + 1.96\sigma/\sqrt{n}$ and 1 in 40 below the limit $\mu - 1.96\sigma/\sqrt{n}$, that is, 1 in 20 outside $\mu \pm 1.96\sigma/\sqrt{n}$. These limits are referred to as the upper and lower 1 in 40 (or 0.025) lines or inner control lines, respectively.

Further, only one subgroup mean in 1000 will lie above the limit $\mu + 3.09\sigma/\sqrt{n}$ and 1 in 1000 below $\mu - 3.09\sigma/\sqrt{n}$, that is, 1 in 500 outside $\mu \pm 3.09\sigma/\sqrt{n}$. These limits are known as the 1 in 1000 (or 0.001) or outer control lines.

To simplify the calculation and the verbal description in practice the multiplying factor 1.96 is rounded to 2 and 3.09 to 3. In some parts of the world, as in the United States, only the $3\sigma/\sqrt{n}$ limits are used. Common practice is to base action on a single point falling beyond the outer limits or successive points outside the inner limit.

The justification for these choices of control limits to indicate a variation that should not be allowed to pass is that practical experience has shown that occurrence of points beyond them is a very useful signal for action to be taken. The correspondence of control limits to explicit probabilities derived from the normal distribution curve may be of value when interpreting the charts, but too much attention should not be paid to the precise probability values associated with the particular multiple of the standard error that is adopted. When quantitative results on quality are required, a more stringent scheme of calculations should be followed, for example, an analysis of variance scheme (Section 4.6) or sequential analysis (Section 3.5). The speed of discovery that something is out of control can be estimated with the following reasoning.

Suppose the average value suddenly shifts away from μ by an amount equivalent to $3\sigma/\sqrt{n}$. Then there is a 1 in 2 chance that the next subgroup average will fall beyond the outer control limit and initiate action. The chance of detection after two readings will be $\frac{1}{2}+(\frac{1}{2}\times\frac{1}{2})$, etc. The average number of tests before the change is detected is termed the average run length. For a $3\sigma/\sqrt{n}$ shift of the mean the average run length is 2. In general the average run length is $1/A$, where A is the tail area of the normal distribution beyond the control line.

Chart for Subgroup Ranges

In this chart the quality parameter precision is plotted and used for control. Now the standard deviation s or the range R of a subgroup of results consisting of n observations is used to construct a control chart. The range R as a rule should be used when the subgroup size $n<13$, as is to be expected in controlling an analytical procedure. The central line for a chart for subgroup ranges is σ_0 or s_0, the estimate of σ_0. The inner and outer limits will be fixed at $D_{0.025}\sigma$ and $D_{0.001}\sigma$, respectively. The same probability levels are used as in the chart for subgroup averages, 0.001 in each tail for the outer limits and 0.025 in each tail for the inner limits.

Note that the lower and upper limits are not symmetric about the average, because the chance that the range will be higher than the average is not equal to the chance that the range will be lower than the average (Table 3.9).

TABLE 3.9. Outer and Inner Limits for Subgroup Ranges

Number of Observations in a Subgroup (n)	Lower Limit		Upper Limit	
	Outer Limit $D_{0.001}$	Inner Limit $D_{0.025}$	Inner Limit $D_{0.975}$	Outer Limit $D_{0.999}$
2	0.00	0.04	3.17	4.65
3	0.06	0.30	3.68	5.06
4	0.20	0.59	3.98	5.31
5	0.37	0.85	4.20	5.48
6	0.54	1.06	4.36	5.62
7	0.69	1.25	4.49	5.73
8	0.83	1.41	4.61	5.82
9	0.96	1.55	4.70	5.90
10	1.08	1.67	4.79	5.97
11	1.20	1.78	4.86	6.04
12	1.30	1.88	4.92	6.09

Cusum Charts

The charts for average and range can be used when sudden changes should be detected. In controlling the quality of analytical methods, this is particularly so when sudden changes in the method, application of wrong procedures, interchange of reagents, and the breakdown of parts of instruments have to be detected. When more gradual changes should be detected, as for instance aging of reagents and obsolence of standard curves, another method of plotting the results should be used. This method is known as the cusum chart.

Here the decision is based on the whole of the recent sequence of measurements. Instead of plotting the subgroup averages $\bar{x}_1, \bar{x}_2, \ldots$, a reference value k is chosen and the cumulative sum is calculated:

$$s_i = \sum_{j=1}^{i} \left(\bar{x}_j - k \right) \tag{3.63}$$

Each sum s_i is obtained from its predecessor by the operation of adding the new difference. k is a reference value that is chosen equal to the average, but calculation of the cumulative sums (abbreviated to cusum) is simplified if a rounded value is chosen. If the mean value of the subgroup averages remains close to the reference value, some of the differences are positive and some negative, so that the cusum chart is essentially horizontal. However, if the average value of the process rises to a new constant level, more of the differences become positive and the mean path slopes upward. The position of the change in the cusum slope indicates the moment the change in the quality parameter occurred. However, the visual picture depends to some extend on the scales chosen for the axes of the chart. If the horizontal distance between the plotted points is regarded as one unit, it is recommended that the same distance on the vertical scale represent approximately $2\sigma/\sqrt{n}$ property units, where σ is the population standard deviation.

The unit distance on the vertical scale does not have to be exactly equivalent to $2\sigma/\sqrt{n}$ units and in fact it eases the task of plotting the cusum, if a suitable rounded lower value is chosen.

Just as action decisions from a Shewhart chart are signaled by points falling outside the control limits, a rule is needed to decide when the change of slope indicates some assignable cause in the method of analysis. One method for taking these decisions is to use a V–shaped mask. This is a graphic aid that can be used to quantify the disturbance in the cusum. The mask can be constructed as a solid truncated V in a material such as cardboard or the V can be engraved in a sheet of transparent perspex. The mask is positioned on the cusum with the axis of the V parallel to the horizontal axis of the chart (Figure 3.13). The point P, at a distance $d=2x$ (the unit length of the horizontal axis of the cusum chart) from T, is superimposed on the most recent cusum value S_i. If one of the preceding cusum S_i's is positioned below the limit of the V, OT, this is an indication of a significant positive change of μ. If the cusum line crosses limb BT, μ has changed in a downward direction. In both circumstances action is undertaken. A V mask with an angle $\theta=30°$ and a lead distance d as indicated is equivalent with $3\sigma/\sqrt{n}$ limits. This has an average run length of 500 points.

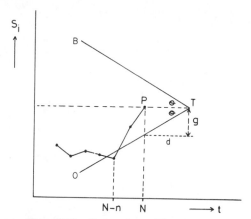

Figure 3.13. Cusum chart with mask.

Sequential Analysis

As might be supposed, sequential analysis (Section 3.5) can be used for quality control of analytical chemical procedures. In fact, the cusum technique is a somewhat simplified variant of sequential analysis. Therefore the cusum chart seems more suitable to the needs of control in a laboratory, for one of the axioms used states that all laboratory workers can evaluate the quality measurements and undertake action when required.

3.4.3   Youden Plot

A very popular and effective controlling device for analytical laboratories is the round-robin test with graphic reporting. The method was described in detail by Youden (YO 72) and is therefore often called the Youden plot method (Youden himself refers to it as the two-sample chart). He demonstrates the role of systematic and random errors or, stated in another way, that of accuracy and precision. The data for these charts (Figure 3.14) are obtained by having two similar materials, designated x and y, analyzed by each collaborator with a request for one determination per sample. Expanded rectangular x and y scales form the representation space for the pair of results provided by each laboratory. The pattern formed by the points conclusively demonstrates the major role played by the various systematic errors. Consider that random errors are really the cause of the scatter. In such a situation the two determinations may err in being both low, both high, x low and y high, or x high and y low. Since random errors

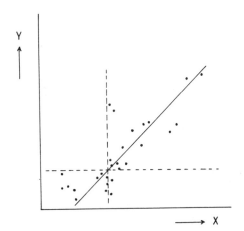

Figure 3.14.   The Youden plot.

are equally likely, all of the four possible results should be equally likely. Thus in every quadrant, formed by subdividing the field around the mean of the points in the representation space, the number of points corresponding to the laboratories should be equal. Otherwise stated, the probability density function consists of concentric circles with the mean as center. If the points are considered to have a normalized Gaussian probability function, the x results are scattered around the mean x_0 with a probability density

$$P_x = \frac{1}{\sigma\sqrt{2\pi}}\exp\left(-\frac{(x-x_0)^2}{2\sigma^2}\right) \qquad (\sigma = \text{standard deviation}) \quad (3.64)$$

The same situation holds for the y results. The standard deviation σ of x and y are the same, because the samples were similar; thus

$$P_y = \frac{1}{\sigma\sqrt{2\pi}}\exp\left(-\frac{(y-y_0)^2}{2\sigma^2}\right) \qquad (3.65)$$

The probability density function of the points (x, y) in the representation space equals

$$P_{x,y} = P_x * P_y = \frac{1}{2\pi\sigma^2}\exp\left\{-\frac{1}{2\sigma^2}\left[(x-x_0)^2 + (y-y_0)^2\right]\right\} \quad (3.66)$$

In Figure 3.15 the lines with equal density for σ and 2σ are shown. As can be seen from eq. 3.66 these lines are concentric circles. Frequently, however, no such distribution has been found. The points are nearly always found dominantly in two quadrants: the $+ +$ upper right quadrant and the $- -$ lower left quadrant. Generally the points form an elliptical pattern with the major axis of the ellipse running diagonally at an angle of 45° to the x-axis, instead of being scattered within a circle. This indicates that, when a laboratory gets a result denoted as "high" (in reference to the consensus) with one material, it is most probable to be similarly high with the other material. The same statement holds for low results. The elongation of the ellipse is a measure for the accuracy of the method, scales along the x- and y-axes being the same. The distance of each point along the 45° axis to the mean is a measure of the systematic error that the laboratory is likely to have been making.

This is in complete agreement with the common understanding that, although good checks are obtained within one laboratory due to good precision, it is more difficult to check the results from interlaboratory

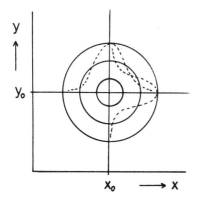

Figure 3.15. The probability density of the Youden plot.

comparisons. Incidentally the second laboratory will also get good checks. Evidently each laboratory has good precision but something operates to displace the results in one laboratory from those obtained in another one. It is appropriate to assume that each laboratory has its own systematic error.

A laboratory that has large systematic errors will be represented by a point x, y in the Youden plot that sticks to the 45° axis, but that lies quite a distance from x_0, y_0. The distance of the point $T(x_1, y_1)$ (Figure 3.16) to the 45° axis is an estimate of the random error (the precision) of this laboratory. The distance from S (the projection of T on the 45° axis) to x_0, y_0 is an estimate of the systematic error.

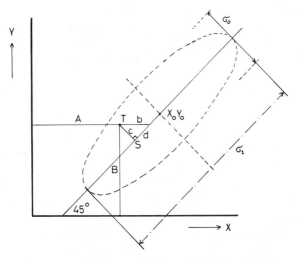

Figure 3.16. Precision and accuracy in the Youden plot.

In fact the Youden plot is a graphic representation of an analysis of variance (Section 4.6). The correlation between these methods can be seen by comparing some parameters.

In an analysis of variance the random error σ_0 is estimated by

$$\sigma_0^2 = \sum_{i=1}^{k} \sum_{j=1}^{n} \frac{(x_{ij} - \bar{x}_i)^2}{k(n-1)} = \frac{\sum_{i=1}^{k} \sum_{j=1}^{n} (x_{ij} - \bar{x}_i)^2}{k} \tag{3.67}$$

with k = number of laboratories
n = number of analyses for each laboratory (here $n = 2$)
In the Youden plot (Figure 3.16)

$$\sum_{i=1}^{k} \sum_{j=1}^{n} (x_{ij} - \bar{x}_i)^2 = \sum_{i=1}^{k} \left\{ \left[A_i - \left(A_i + \tfrac{1}{2}b_i \right) \right]^2 + \left[B_i - \left(B_i - \tfrac{1}{2}b_i \right) \right]^2 \right\}$$

$$= \sum_{i=1}^{k} \left[\left(\tfrac{1}{2}b_i \right)^2 + \left(\tfrac{1}{2}b_i \right)^2 \right]$$

$$= \sum_{i=1}^{k} \tfrac{1}{2}b_i^2 = \sum_{i=1}^{k} c_i^2 \tag{3.68}$$

$$\rightarrow \sigma_0^2 = \frac{\sum_{i=1}^{k} c_i^2}{k(n-1)} \tag{3.69}$$

In this case $n = 2$, so

$$\sigma_0^2 = \frac{\sum_{i=1}^{k} c_i^2}{k} \tag{3.70}$$

In the analysis of variance the total error $\sigma_0^2 + n\sigma_1^2$ is estimated by

$$\sigma_0^2 + n\sigma_1^2 = n \frac{\sum_{i=1}^{k} (\bar{x}_i - \bar{x})^2}{k-1} \tag{3.71}$$

With $n = 2$

$$\sigma_0^2 + 2\sigma_1^2 = 2\frac{\sum\limits_{i=1}^{k}(\bar{x}_i - \bar{x})^2}{k-1} \tag{3.72}$$

In the Youden plot,

$$\sigma_0^2 + 2\sigma_1^2 = 2\frac{\sum\limits_{i=1}^{k} d_i^2}{k-1} \tag{3.73}$$

The significance of the difference between total error and precision σ_0 can be tested by applying the F-test:

$$F = \frac{2\left(\sum\limits_{i=1}^{k} d_i^2\right)/(k-1)}{\left(\sum\limits_{i=1}^{k} c_i^2\right)/k} \tag{3.74}$$

For the value of $F >$ the tabulated F (for a reliability α, $\nu_1 = k - 1$ and $\nu_2 = k$) (Table 3.10) it may be supposed that a systematic error exists.

$$\sigma_1^2 = \left[\frac{\sum\limits_{i=1}^{k} d_i^2}{k-1}\right] - \left[\frac{\frac{1}{2}\left(\sum\limits_{i=1}^{k} c_i^2\right)}{k}\right] \tag{3.75}$$

The Youden plot not only can be used for the comparison of the quality of laboratories with regard to their results, but also for the comparison of the quality of two methods of analysis with regard to their (in)sensitivity to various laboratories. When comparing laboratories it is important that a sufficient number of them will be involved, because only in such situations can quantitative results be obtained.

In practice, however, often only the qualitative result is presented in graphic form. It should be noted that this works quite well because "outliers," laboratories having points far from x_0, y_0 along the 45° axis, seldom appear at the same spot when the round-robin test is repeated.

It must be emphasized that quantitative results can be obtained only when the distribution of the data is not significantly different from a

TABLE 3.10 Probability Points of the Variance Ratio (F-Distribution)

ϕ_N (corresponding to greater mean square)

ϕ_D	Probability Point	1	2	3	4	5	6	7	8	9	10	12	15	20	24	30	40	60	120	∞
1	0·100	39·9	49·5	53·6	55·8	57·2	58·2	58·9	59·4	59·9	60·2	60·7	61·2	61·7	62·0	62·3	62·5	62·8	63·1	63·3
	0·050	161	199	216	225	230	234	237	239	241	242	244	246	248	249	250	251	252	253	254
	0·025	648	800	864	900	922	937	948	957	963	969	977	985	993	997	1001	1006	1010	1014	1018
	0·010	4052	4999	5403	5625	5764	5859	5928	5982	6022	6056	6106	6157	6209	6235	6261	6287	6313	6339	6366
2	0·100	8·53	9·00	9·16	9·24	9·29	9·33	9·35	9·37	9·38	9·39	9·41	9·42	9·44	9·45	9·46	9·47	9·47	9·48	9·49
	0·050	18·5	19·0	19·2	19·2	19·3	19·3	19·4	19·4	19·4	19·4	19·4	19·4	19·4	19·5	19·5	19·5	19·5	19·5	19·5
	0·025	38·5	39·0	39·2	39·2	39·3	39·3	39·4	39·4	39·4	39·4	39·4	39·4	39·4	39·5	39·5	39·5	39·5	39·5	39·5
	0·010	98·5	99·0	99·2	99·2	99·3	99·3	99·4	99·4	99·4	99·4	99·4	99·4	99·4	99·5	99·5	99·5	99·5	99·5	99·5
3	0·100	5·54	5·46	5·39	5·34	5·31	5·28	5·27	5·25	5·24	5·23	5·22	5·20	5·18	5·18	5·17	5·16	5·15	5·14	5·13
	0·050	10·1	9·55	9·28	9·12	9·01	8·94	8·89	8·85	8·81	8·79	8·74	8·70	8·66	8·64	8·62	8·59	8·57	8·55	8·53
	0·025	17·4	16·0	15·4	15·1	14·9	14·7	14·6	14·5	14·5	14·4	14·3	14·3	14·2	14·1	14·1	14·0	14·0	13·9	13·9
	0·010	34·1	30·8	29·5	28·7	28·2	27·9	27·7	27·5	27·3	27·2	27·1	26·9	26·7	26·6	26·5	26·4	26·3	26·2	26·1
4	0·100	4·54	4·32	4·19	4·11	4·05	4·01	3·98	3·95	3·94	3·92	3·90	3·87	3·84	3·83	3·82	3·80	3·79	3·78	3·76
	0·050	7·71	6·94	6·59	6·39	6·26	6·16	6·09	6·04	6·00	5·96	5·91	5·86	5·80	5·77	5·75	5·72	5·69	5·66	5·63
	0·025	12·2	10·6	10·0	9·60	9·36	9·20	9·07	8·98	8·90	8·84	8·75	8·66	8·56	8·51	8·46	8·41	8·36	8·31	8·26
	0·010	21·2	18·0	16·7	16·0	15·5	15·2	15·0	14·8	14·7	14·5	14·4	14·2	14·0	13·9	13·8	13·7	13·7	13·6	13·5
5	0·100	4·06	3·78	3·62	3·52	3·45	3·40	3·37	3·34	3·32	3·30	3·27	3·24	3·21	3·19	3·17	3·16	3·14	3·12	3·10
	0·050	6·61	5·79	5·41	5·19	5·05	4·95	4·88	4·82	4·77	4·74	4·68	4·62	4·56	4·53	4·50	4·46	4·43	4·40	4·36
	0·025	10·0	8·43	7·76	7·39	7·15	6·98	6·85	6·76	6·68	6·62	6·52	6·43	6·33	6·28	6·23	6·18	6·12	6·07	6·02
	0·010	16·3	13·3	12·1	11·4	11·0	10·7	10·5	10·3	10·2	10·1	9·89	9·72	9·55	9·47	9·38	9·29	9·20	9·11	9·02
6	0·100	3·78	3·46	3·29	3·18	3·11	3·05	3·01	2·98	2·96	2·94	2·90	2·87	2·84	2·82	2·80	2·78	2·76	2·74	2·72
	0·050	5·99	5·14	4·76	4·53	4·39	4·28	4·21	4·15	4·10	4·06	4·00	3·94	3·87	3·84	3·81	3·77	3·74	3·70	3·67
	0·025	8·81	7·26	6·60	6·23	5·99	5·82	5·70	5·60	5·52	5·46	5·37	5·27	5·17	5·12	5·07	5·01	4·96	4·90	4·85
	0·010	13·7	10·9	9·78	9·15	8·75	8·47	8·26	8·10	7·98	7·87	7·72	7·56	7·40	7·31	7·23	7·14	7·06	6·97	6·88
7	0·100	3·59	3·26	3·07	2·96	2·88	2·83	2·78	2·75	2·72	2·70	2·67	2·63	2·59	2·58	2·56	2·54	2·51	2·49	2·47
	0·050	5·59	4·74	4·35	4·12	3·97	3·87	3·79	3·73	3·68	3·64	3·57	3·51	3·44	3·41	3·38	3·34	3·30	3·27	3·23
	0·025	8·07	6·54	5·89	5·52	5·29	5·12	4·99	4·90	4·82	4·76	4·67	4·57	4·47	4·42	4·36	4·31	4·25	4·20	4·14
	0·010	12·2	9·55	8·45	7·85	7·46	7·19	6·99	6·84	6·72	6·62	6·47	6·31	6·16	6·07	5·99	5·91	5·82	5·74	5·65

TABLE 3.10 (*continued*)

8	0·100	3·46	3·11	2·92	2·81	2·73	2·67	2·62	2·59	2·56	2·54	2·50	2·46	2·42	2·40	2·38	2·36	2·34	2·32	2·29	0·100	8
	0·050	5·32	4·46	4·07	3·84	3·69	3·58	3·50	3·44	3·39	3·35	3·28	3·22	3·15	3·12	3·08	3·04	3·01	2·97	2·93	0·050	
	0·025	7·57	6·06	5·42	5·05	4·82	4·65	4·53	4·43	4·36	4·30	4·20	4·10	4·00	3·95	3·89	3·84	3·78	3·73	3·67	0·025	
	0·010	11·3	8·65	7·59	7·01	6·63	6·37	6·18	6·03	5·91	5·81	5·67	5·52	5·36	5·28	5·20	5·12	5·03	4·95	4·86	0·010	
9	0·100	3·36	3·01	2·81	2·69	2·61	2·55	2·51	2·47	2·44	2·42	2·38	2·34	2·30	2·28	2·25	2·23	2·21	2·18	2·16	0·100	9
	0·050	5·12	4·26	3·86	3·63	3·48	3·37	3·29	3·23	3·18	3·14	3·07	3·01	2·94	2·90	2·86	2·83	2·79	2·75	2·71	0·050	
	0·025	7·21	5·71	5·08	4·72	4·48	4·32	4·20	4·10	4·03	3·96	3·87	3·77	3·67	3·61	3·56	3·51	3·45	3·39	3·33	0·025	
	0·010	10·6	8·02	6·99	6·42	6·06	5·80	5·61	5·47	5·35	5·26	5·11	4·96	4·81	4·73	4·65	4·57	4·48	4·40	4·31	0·010	
10	0·100	3·28	2·92	2·73	2·61	2·52	2·46	2·41	2·38	2·35	2·32	2·28	2·24	2·20	2·18	2·16	2·13	2·11	2·08	2·06	0·100	10
	0·050	4·96	4·10	3·71	3·48	3·33	3·22	3·14	3·07	3·02	2·98	2·91	2·84	2·77	2·74	2·70	2·66	2·62	2·58	2·54	0·050	
	0·025	6·94	5·46	4·83	4·47	4·24	4·07	3·95	3·85	3·78	3·72	3·62	3·52	3·42	3·37	3·31	3·26	3·20	3·14	3·08	0·025	
	0·010	10·0	7·56	6·55	5·99	5·64	5·39	5·20	5·06	4·94	4·85	4·71	4·56	4·41	4·33	4·25	4·17	4·08	4·00	3·91	0·010	
12	0·100	3·18	2·81	2·61	2·48	2·39	2·33	2·28	2·24	2·21	2·19	2·15	2·10	2·06	2·04	2·01	1·99	1·96	1·93	1·90	0·100	12
	0·050	4·75	3·89	3·49	3·26	3·11	3·00	2·91	2·85	2·80	2·75	2·69	2·62	2·54	2·51	2·47	2·43	2·38	2·34	2·30	0·050	
	0·025	6·55	5·10	4·47	4·12	3·89	3·73	3·61	3·51	3·44	3·37	3·28	3·18	3·07	3·02	2·96	2·91	2·85	2·79	2·72	0·025	
	0·010	9·33	6·93	5·95	5·41	5·06	4·82	4·64	4·50	4·39	4·30	4·16	4·01	3·86	3·78	3·70	3·62	3·54	3·45	3·36	0·010	
15	0·100	3·07	2·70	2·49	2·36	2·27	2·21	2·16	2·12	2·09	2·06	2·02	1·97	1·92	1·90	1·87	1·85	1·82	1·79	1·76	0·100	15
	0·050	4·54	3·68	3·29	3·06	2·90	2·79	2·71	2·64	2·59	2·54	2·48	2·40	2·33	2·29	2·25	2·20	2·16	2·11	2·07	0·050	
	0·025	6·20	4·77	4·15	3·80	3·58	3·41	3·29	3·20	3·12	3·06	2·96	2·86	2·76	2·70	2·64	2·59	2·52	2·46	2·40	0·025	
	0·010	8·68	6·36	5·42	4·89	4·56	4·32	4·14	4·00	3·89	3·80	3·67	3·52	3·37	3·29	3·21	3·13	3·05	2·96	2·87	0·010	
20	0·100	2·97	2·59	2·38	2·25	2·16	2·09	2·04	2·00	1·96	1·94	1·89	1·84	1·79	1·77	1·74	1·71	1·68	1·64	1·61	0·100	20
	0·050	4·35	3·49	3·10	2·87	2·71	2·60	2·51	2·45	2·39	2·35	2·28	2·20	2·12	2·08	2·04	1·99	1·95	1·90	1·84	0·050	
	0·025	5·87	4·46	3·86	3·51	3·29	3·13	3·01	2·91	2·84	2·77	2·68	2·57	2·46	2·41	2·35	2·29	2·22	2·16	2·09	0·025	
	0·010	8·10	5·85	4·94	4·43	4·10	3·87	3·70	3·56	3·46	3·37	3·23	3·09	2·94	2·86	2·78	2·69	2·61	2·52	2·42	0·010	

TABLE 3.10 (*continued*)

	p																					p	
24	0·100	2·93	2·54	2·33	2·19	2·10	2·04	1·98	1·94	1·91	1·88	1·83	1·78	1·73	1·70	1·67	1·64	1·61	1·57	1·53	0·100	24
	0·050	4·26	3·40	3·01	2·78	2·62	2·51	2·42	2·36	2·30	2·25	2·18	2·11	2·03	1·98	1·94	1·89	1·84	1·79	1·73	0·050	
	0·025	5·72	4·32	3·72	3·38	3·15	2·99	2·87	2·78	2·70	2·64	2·54	2·44	2·33	2·27	2·21	2·15	2·08	2·01	1·94	0·025	
	0·010	7·82	5·61	4·72	4·22	3·90	3·67	3·50	3·36	3·26	3·17	3·03	2·89	2·74	2·66	2·58	2·49	2·40	2·31	2·21	0·010	
30	0·100	2·88	2·49	2·28	2·14	2·05	1·98	1·93	1·88	1·85	1·82	1·77	1·72	1·67	1·64	1·61	1·57	1·54	1·50	1·46	0·100	30
	0·050	4·17	3·32	2·92	2·69	2·53	2·42	2·33	2·27	2·21	2·16	2·09	2·01	1·93	1·89	1·84	1·79	1·74	1·68	1·62	0·050	
	0·025	5·57	4·18	3·59	3·25	3·03	2·87	2·75	2·65	2·57	2·51	2·41	2·31	2·20	2·14	2·07	2·01	1·94	1·87	1·79	0·025	
	0·010	7·56	5·39	4·51	4·02	3·70	3·47	3·30	3·17	3·07	2·98	2·84	2·70	2·55	2·47	2·39	2·30	2·21	2·11	2·01	0·010	
40	0·100	2·84	2·44	2·23	2·09	2·00	1·93	1·87	1·83	1·79	1·76	1·71	1·66	1·61	1·57	1·54	1·51	1·47	1·42	1·38	0·100	40
	0·050	4·08	3·23	2·84	2·61	2·45	2·34	2·25	2·18	2·12	2·08	2·00	1·92	1·84	1·79	1·74	1·69	1·64	1·58	1·51	0·050	
	0·025	5·42	4·05	3·46	3·13	2·90	2·74	2·62	2·53	2·45	2·39	2·29	2·18	2·07	2·01	1·94	1·88	1·80	1·72	1·64	0·025	
	0·010	7·31	5·18	4·31	3·83	3·51	3·29	3·12	2·99	2·89	2·80	2·66	2·52	2·37	2·29	2·20	2·11	2·02	1·92	1·80	0·010	
60	0·100	2·79	2·39	2·18	2·04	1·95	1·87	1·82	1·77	1·74	1·71	1·66	1·60	1·54	1·51	1·48	1·44	1·40	1·35	1·29	0·100	60
	0·050	4·00	3·15	2·76	2·53	2·37	2·25	2·17	2·10	2·04	1·99	1·92	1·84	1·75	1·70	1·65	1·59	1·53	1·47	1·39	0·050	
	0·025	5·29	3·93	3·34	3·01	2·79	2·63	2·51	2·41	2·33	2·27	2·17	2·06	1·94	1·88	1·82	1·74	1·67	1·58	1·48	0·025	
	0·010	7·08	4·98	4·13	3·65	3·34	3·12	2·95	2·82	2·72	2·63	2·50	2·35	2·20	2·12	2·03	1·94	1·84	1·73	1·60	0·010	
120	0·100	2·75	2·35	2·13	1·99	1·90	1·82	1·77	1·72	1·68	1·65	1·60	1·54	1·48	1·45	1·41	1·37	1·32	1·26	1·19	0·100	120
	0·050	3·92	3·07	2·68	2·45	2·29	2·18	2·09	2·02	1·96	1·91	1·83	1·75	1·66	1·61	1·55	1·50	1·43	1·35	1·25	0·050	
	0·025	5·15	3·80	3·23	2·89	2·67	2·52	2·39	2·30	2·22	2·16	2·05	1·94	1·82	1·76	1·69	1·61	1·53	1·43	1·31	0·025	
	0·010	6·85	4·79	3·95	3·48	3·17	2·96	2·79	2·66	2·56	2·47	2·34	2·19	2·03	1·95	1·86	1·76	1·66	1·53	1·38	0·010	
∞	0·100	2·71	2·30	2·08	1·94	1·85	1·77	1·72	1·67	1·63	1·60	1·55	1·49	1·42	1·38	1·34	1·30	1·24	1·17	1·00	0·100	∞
	0·050	3·84	3·00	2·60	2·37	2·21	2·10	2·01	1·94	1·88	1·83	1·75	1·67	1·57	1·52	1·46	1·39	1·32	1·22	1·00	0·050	
	0·025	5·02	3·69	3·12	2·79	2·57	2·41	2·29	2·19	2·11	2·05	1·94	1·83	1·71	1·64	1·57	1·48	1·39	1·27	1·00	0·025	
	0·010	6·63	4·61	3·78	3·32	3·02	2·80	2·64	2·51	2·41	2·32	2·18	2·04	1·88	1·79	1·70	1·59	1·47	1·32	1·00	0·010	

119

normal one. For a skewed distribution, as may be the case when the samples x and y are near the limit of detection, an appropriate linearizing method should be applied (Section 4.1.2), or the quantitative presentation should be abandoned in favor of the graphic plot.

3.4.4   Ranking Test (YO 72)

Another way to assess the quality of a laboratory or an analytical method is the application of the ranking test. The data from a collaborative test may be arranged in a classification scheme as shown in Table 3.11. The right-hand side of the table shows the data substituted by rankings assigned to the laboratories according to the measured concentration for a series of samples. Rank 1 is given to the largest amount, rank 2 to the next largest, and so on. When two laboratories have the same score for the xth place, each laboratory is assigned the rank $x + \frac{1}{2}$. For a triple tie for the xth place, all these get the rank $x + 1$. The laboratory score equals the sum of the ranks it receives.

$$M = \sum_{i=1}^{n} R_i \qquad (3.76)$$

with R_i = ranking for sample i
$\quad n$ = number of quality parameters

TABLE 3.11.   The Ranking Test

Coll. No.	Results (%) Samples					Ranked Results Samples					Coll. Score
	1	2	3	4	5	1	2	3	4	5	
7	4.59	1.46	5.64	2.19	27.32	9	5.5	6	4	3	27.5
8	4.94	1.52	5.68	2.28	26.44	1	1	3	2	10	17
9	4.80	1.40	5.62	2.12	26.89	3.5	8.5	7.5	6.5	8	34
10	4.73	1.46	5.65	2.09	27.17	5	5.5	5	8	4	27.5
11	4.72	1.51	5.62	2.12	27.00	6.5	2.5	7.5	6.5	6	29
12	4.80	1.51	5.80	2.29	27.48	3.5	2.5	1	1	1	9[a]
13	4.45	1.40	5.45	2.07	27.02	10	8.5	10	9	5	42.5
15	4.72	1.50	5.58	2.27	26.76	6.5	4	9	3	9	31.5
16	4.63	1.32	5.69	2.04	26.92	8	10	2	10	7	37
17	4.88	1.42	5.67	2.16	27.39	2	7	4	5	2	20

[a] Designates unusually low score.

For n samples, the minimum score is n and the maximum score is nm, where m denotes the number of participating laboratories. A laboratory that scores the highest amount on every one of the n materials gets the score of n. Such a score is obviously associated with a laboratory that consistently gets high results, and the presumption is that this laboratory has a pronounced systematic error. When the results of the laboratories stem from a random process, the scores tend to cluster around the value $n(m+1)/2$. The statistical distribution of such scores has been tabulated in Table 3.12 (YO 72).

TABLE 3.12. Approximate 5% Two-Tail Limits for Ranking Scores

No. of Labs.	Number of Materials												
	3	4	5	6	7	8	9	10	11	12	13	14	15
3		4	5	7	8	10	12	13	15	17	19	20	22
		12	15	17	20	22	24	27	29	31	33	36	38
4		4	6	8	10	12	14	16	18	20	22	24	26
		16	19	22	25	28	31	34	37	40	43	46	49
5		5	7	9	11	13	16	18	21	23	26	28	31
		19	23	27	31	35	38	42	45	49	52	56	59
6	3	5	7	10	12	15	18	21	23	26	29	32	35
	18	23	28	32	37	41	45	49	54	58	62	66	70
7	3	5	8	11	14	17	20	23	26	29	32	36	39
	21	27	32	37	42	47	52	57	62	67	72	76	81
8	3	6	9	12	15	18	22	25	29	32	36	39	43
	24	30	36	42	48	54	59	65	70	76	81	87	92
9	3	6	9	13	16	20	24	27	31	35	39	43	47
	27	34	41	47	54	60	66	73	79	85	91	97	103
10	4	7	10	14	17	21	26	30	34	38	43	47	51
	29	37	45	52	60	67	73	80	87	94	100	107	114
11	4	7	11	15	19	23	27	32	36	41	46	51	55
	32	41	49	57	65	73	81	88	96	103	110	117	125
12	4	7	11	15	20	24	29	34	39	44	49	54	59
	35	45	54	63	71	80	88	96	104	112	120	128	136
13	4	8	12	16	21	26	31	36	42	47	52	58	63
	38	48	58	68	77	86	95	104	112	121	130	138	147
14	4	8	12	17	22	27	33	38	44	50	56	61	67
	41	52	63	73	83	93	102	112	121	130	139	149	158
15	4	8	13	18	23	29	35	41	47	53	59	65	71
	44	56	67	78	89	99	109	119	129	139	149	159	169

Reprinted with permission from W. J. Youden and E. H. Steiner, *Statistical Manual of the AOAC*, Association of Official Analytical Chemists, Washington, D.C., 1975.

The upper and lower limits shown in this table have been determined to be based on the assumption that there is a 95% reliability that the rankings between these limits result from a random process. In extreme situations most of the laboratories may get approximately the same ranking on each material. Here the scores approximate the values $n, 2n, \ldots, mn$. Obviously this is an indictment of the analytical method: presumably either its description is inadequately written or its implication is unacceptably sensitive to the environmental differences encountered in the various laboratories.

3.5 SEQUENTIAL ANALYSIS

When performing a series of experiments, one sometimes has to decide whether the experiment gives a result that confirms the hypothesis that is tested or whether, alternatively, this hypothesis must be rejected. This testing can be done by applying statistics to the set of observations. There are then three possible decisions:

1. From the obtained results it is not yet possible to reject the hypothesis nor to accept it.
2. The hypothesis is rejected.
3. The hypothesis is accepted.

With the use of the proper statistics decisions 2 and 3 can be made with a certain degree of confidence. When the experiment has been planned without prior knowledge of the statistical behavior of the experiment, it is possible that decisions 2 and 3 will be made with a confidence level that is much higher than required for the problem under investigation. The conclusion should be that the number of observations could have been less. If, however, decision 1 is the result, this means that the experiment has to be continued. The question to be answered now is, how long should the experiment be continued in order to reach decisions 2 or 3 with sufficient—but not too much—confidence?

It is obvious that the design of an experiment consisting of a series of observations is more critical when the cost—in money or time—per observation is high.

A method that circumvents the problems just described is sequential analysis. By this method a test is conducted that allows a decision to be made after each observation. Thus the number of observations is minimal. The theory of this method is described in many textbooks (e.g., DA 71, DA 72, WA 74). Application to various problems requires a thorough

TABLE 3.13. Comparison of Sequential and Nonsequential Analyses

Sequential Analysis	Nonsequential Analysis
The number of observations required is not known in advance	The number of observations is known before the experiment starts
After each observation one out of three possible decisions is selected: 1. Continue the experiment and make a new observation 2. Reject the null hypothesis and terminate the experiment 3. Accept the null hypothesis and terminate the experiment	After the experiment, comprising a series of observations, three decisions are possible: 1. Start a new series of observations 2. Reject the null hypothesis 3. Accept the null hypothesis
After each observation the decision to terminate the experiment may result The decision to terminate the experiment as a rule results in fewer observations than in nonsequential analysis	The experiment is terminated at the end of the series of observations
The method is recommended when the measuring time is short compared with computing time The method is preferred for expensive measurements	The method is recommended when the measuring time is long compared with computing time

knowledge of statistics. Here we describe only two cases, problems that are often encountered in analytical chemistry.

Sequential analysis is a method used to guide statistical interference in observation series in order to limit the number of observations of that series, which is not known in advance. The decision whether to terminate the experiment depends on the result of the observations collected so far. A comparison of sequential and nonsequential analysis is given in Table 3.13. In a sequential test the number of observations required depends on the result of the observations and therefore is not fixed in advance. After each observation a decision is made whether the null hypothesis H_0 will be accepted or rejected, or if the experiment will be continued. The decision to terminate the experiment at any moment depends on the observations made so far.

3.5.1 Mathematical Description

Suppose a hypothesis H has to be tested by performing an experiment that results in the variable x. $x_1, x_2, \ldots, x_m$ are m observations of the variable x.

The progress of the test can be shown graphically by computing the value of a function of all observations, $F(m)$. This value is plotted as a function of the number of observations. The sample space M_m can be subdivided into three mutually exclusive subspaces:

R_m^0:   space where H is accepted
R_m':   space where H is rejected
R_m: space where no decision is possible

In Figure 3.17 two boundaries are indicated. The direction of these boundaries depends on the values of α and β and the parameters tested. (α = probability of an error of the first kind: reject H wrongly. β = probability of an error of the second kind: accept H wrongly.) When the function value crosses one of the boundaries and enters zone II or III, the experiment is terminated by the decision dictated by the zone number.

Example

Suppose a random variable x with a probability distribution $f(x)$ depends on one unknown parameter θ: $\{f(x, \theta)\}$. The null hypothesis reads H_0: $\theta = \theta_0$. The alternative is H_1: $\theta = \theta_1$. If H_0 is accepted, the distribution of x is $f(x, \theta_0)$; if H_1 is accepted it is $f(x, \theta_1)$. With every positive integer m the probability that a series of observations $x_1, x_2, \ldots, x_m$ results is

$$P_{0, m} = f(x_1, \theta_0), \ldots, f(x_m, \theta_0) \text{ when } H_0 \text{ is accepted}$$

$$P_{1, m} = f(x_1, \theta_1), \ldots, f(x_m, \theta_1) \text{ when } H_1 \text{ is accepted}$$

To test whether the series of observations has to be continued the ratio $P_{1, m}/P_{0, m}$

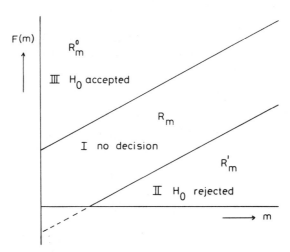

Figure 3.17. Sequential analysis, decision areas.

is computed after each measurement and compared with two constants A and B. These constants are defined as

$$A = \frac{1-\beta}{\alpha} \quad \text{and} \quad B = \frac{\beta}{1-\alpha}$$

Now the three possible situations are as follows:

$$\text{I.} \quad B < \frac{P_{1,m}}{P_{0,m}} < A \tag{3.77}$$

The series of observations is extended with an observation x_{m+1}.

$$\text{II.} \quad \frac{P_{1,m}}{P_{0,m}} \geqslant A \tag{3.78}$$

The experiment is terminated with rejection of H_0 and acceptance of H_1.

$$\text{III.} \quad \frac{P_{1,m}}{P_{0,m}} \leqslant B \tag{3.79}$$

The experiment is terminated with acceptance of H_0 and rejection of H_1.

Often eq. 3.77 is logarithmized thus:

$$\log B < \log\left(\frac{P_{1,m}}{P_{0,m}}\right) < \log A \tag{3.80}$$

in which

$$\log\left(\frac{P_{1,m}}{P_{0,m}}\right) = \log\frac{f(x_1\theta_1),\ldots,f(x_m\theta_1)}{f(x_1\theta_0),\ldots,f(x_m\theta_0)} =$$

$$= \log\frac{f(x_1\theta_1)}{f(x_1\theta_0)} + \log\frac{f(x_2\theta_1)}{f(x_2\theta_0)} + \cdots + \log\frac{f(x_m\theta_1)}{f(x_m\theta_0)} \tag{3.81}$$

Substituting z_i according to

$$z_i = \log\frac{f(x_i\theta_1)}{f(x_i\theta_0)} \tag{3.82}$$

results in

$$\log B < \sum_{i=1}^{m} z_i < \log A \tag{3.83}$$

An example is the test for accuracy, that is, whether a method of analysis has the same mean as a standard method, or is higher or lower. A normal distribution with standard deviation σ for both methods is assumed, the mean of the standard method being x_0. A bias Δ must be detected with given confidence α and β. This is a combination of two tests: a test on a deviation $> +\Delta$ and a test on a deviation $> |-\Delta|$. Therefore the value of α, the probability that the null hypothesis will be rejected, will be halved; thus $\alpha' = \alpha/2$.

A. The first test is on a positive deviation of more than Δ. The null hypothesis is

$$H_0: \qquad x = x_0 = \mu$$
$$H_1: \qquad x = x_1 = x_0 + \Delta$$

The experiment is continued when $x_0 < x < x_1$.

Since we assumed a Gaussian distribution, the probability that a series of m observations leads to acceptance of H_0 reads

$$P_{0,m} = k\exp\left(-\frac{1}{2\sigma^2}\sum_{q=1}^{m}(x_q - x_0)^2\right) \qquad (3.84)$$

where $\sigma =$ the standard deviation
$\quad\quad k =$ constant
The probability that H_0 is rejected is

$$P_{1,m} = k\exp\left(-\frac{1}{2\sigma^2}\sum_{q=1}^{m}(x_q - x_1)^2\right) \qquad (3.85)$$

$$\frac{P_{1,m}}{P_{0,m}} = \frac{\exp\left[-(1/2\sigma^2)\sum\limits_{q=1}^{m}(x_q - x_1)^2\right]}{\exp\left[-(1/2\sigma^2)\sum\limits_{q=1}^{m}(x_q - x_0)^2\right]}$$

$$= \exp\frac{1}{2\sigma^2}\left[-\sum_{q=1}^{m}(x_q - x_1)^2 + \sum_{q=1}^{m}(x_q - x_0)^2\right]$$

$$= \exp\frac{1}{2\sigma^2}\left[2(x_1 - x_0)\sum_{q=1}^{m}x_q - m(x_1^2 - x_0^2)\right]$$

$$= \exp\frac{\Delta}{2\sigma^2}\left[2\sum_{q=1}^{m}x_q - m(2x_0 + \Delta)\right] \qquad (3.86)$$

where $\Delta = x_1 - x_0$.

$$\ln\frac{P_{1,m}}{P_{0,m}} = \frac{\Delta}{2\sigma^2}\left[2\sum_{q=1}^{m}x_q - m(2x_0 + \Delta)\right]$$

$$= \frac{\Delta}{\sigma^2}\left[\sum_{q=1}^{m}x_q - m\left(x_0 + \tfrac{1}{2}\Delta\right)\right] \qquad (3.87)$$

A1. The experiment is continued if (by eq. 3.83)

$$\ln B < \frac{\Delta}{\sigma^2} \sum_{q=1}^{m} x_q - \frac{m\Delta}{\sigma^2}\left(x_0 + \tfrac{1}{2}\Delta\right) < \ln A \qquad (3.88)$$

or

$$\frac{\sigma^2}{\Delta}\ln B + m\left(x_0 + \tfrac{1}{2}\Delta\right) < \sum_{q=1}^{m} x_q < \frac{\sigma^2}{\Delta}\ln A + m\left(x_0 + \tfrac{1}{2}\Delta\right) \qquad (3.89)$$

AII. The experiment is terminated and H_0 rejected if

$$\sum_{q=1}^{m} x_q \geqslant \frac{\sigma^2}{\Delta}\ln A + m\left(x_0 + \tfrac{1}{2}\Delta\right) \qquad (3.90)$$

AIII. The experiment is terminated and H_0 accepted if

$$\sum_{q=1}^{m} x_q \leqslant \frac{\sigma^2}{\Delta}\ln B + m\left(x_0 + \tfrac{1}{2}\Delta\right) \qquad (3.91)$$

B. Analogous reasoning for the negative deviation leads to

$$H_0: \qquad x = x_0$$

$$H_1: \qquad x = x_1 = x_0 - \Delta$$

BI. The experiment is continued if

$$-\frac{\sigma^2}{\Delta}\ln A + m\left(x_0 - \tfrac{1}{2}\Delta\right) < \sum_{q=1}^{m} x_q < -\frac{\sigma^2}{\Delta}\ln B + m\left(x_0 - \tfrac{1}{2}\Delta\right) \quad (3.92)$$

BII. The experiment is terminated and H_0 rejected if

$$\sum_{q=1}^{m} x_q \leqslant -\frac{\sigma^2}{\Delta}\ln A + m\left(x_0 - \tfrac{1}{2}\Delta\right) \qquad (3.93)$$

BIII. The experiment is terminated and H_0 accepted if

$$\sum_{q=1}^{m} x_q \geqslant -\frac{\sigma^2}{\Delta}\ln B + m\left(x_0 - \tfrac{1}{2}\Delta\right) \qquad (3.94)$$

Combination of situations A and B can be visualized by plotting $\sum_{q=1}^{m} x_q$ as a function of m (Figure 3.18). Both decisions AI and BI lead to a continuation of the

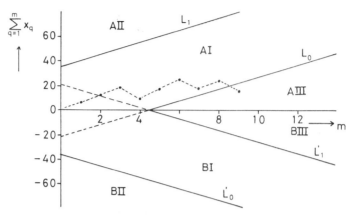

Figure 3.18. Sequential analysis of the accuracy of a series of observations (DA 71).

experiment. The boundaries L_0, L_1, L_0', and L_1' are given by, respectively

$$L_0 = \frac{\sigma^2}{\Delta} \ln B + m\left(x_0 + \tfrac{1}{2}\Delta\right) \text{ between decisions AI/AIII} \qquad (3.95)$$

$$L_1 = \frac{\sigma^2}{\Delta} \ln A + m\left(x_0 + \tfrac{1}{2}\Delta\right) \text{ between decisions AI/AII} \qquad (3.96)$$

$$L_0' = -\frac{\sigma^2}{\Delta} \ln A + m\left(x_0 - \tfrac{1}{2}\Delta\right) \text{ between decisions BI/BII} \qquad (3.97)$$

$$L_1' = -\frac{\sigma^2}{\Delta} \ln B + m\left(x_0 - \tfrac{1}{2}\Delta\right) \text{ between decisions BI/BIII} \qquad (3.98)$$

TABLE 3.14. Progress of a Sequential Analysis of Accuracy

No. of Observation (m)	Observation x_q	Sum $\sum_{q=1}^{m} x_q$	Decision	Conclusion[a] H_0	Conclusion[a] H_0'	Result
1	7	7	AI, BI	—	—	Continue
2	5	12	AI, BI	—	—	Continue
3	8	20	AI, BIII	—	a	Continue
4	−11	9	AI	—		Continue
5	10	19	AI	—		Continue
6	8	27	AI	—		Continue
7	−9	18	AI	—		Continue
8	6	24	AI	—		Continue
9	−7	17	AIII	a		Stop

[a] a = accepted; — = neither accepted or rejected.
No significant bias detected.

An experiment was conducted in which the null hypothesis stated that the mean of the method should be equal to the mean of the standard method. A bias of $\Delta > 10$ should be detected with $\alpha = 0.05$ and $\beta = 0.10$. The mean of the standard method $x_0 = 0$, $\sigma = 10$. Now

$$\ln A = \ln \frac{1-\beta}{\alpha'} = \ln \frac{0.90}{0.025} = 3.58$$

$$\ln B = \ln \frac{\beta}{1-\alpha'} = \ln \frac{0.10}{0.975} = -2.28$$

This leads to the boundaries (Figure 3.18)

$$L_0 = -22.8 + 5m$$

$$L_1 = \quad 35.8 + 5m$$

$$L_0' = -35.8 - 5m$$

$$L_1' = \quad 22.8 - 5m$$

A series of observations as given in Table 3.14 can then be plotted (Figure 3.18).

C.   A test on the precision of a method can also be conducted by sequential analysis. Now the null hypothesis is that the standard deviation σ of the method is equal to the standard deviation of a standard method, σ_0. The errors that are probable in the test are again α and β. The null hypothesis is rejected if $\sigma = \sigma_1 = \sigma_0 + \delta$.

$$H_0: \quad \sigma = \sigma_0$$

$$H_1: \quad \sigma = \sigma_1 = \sigma_0 + \delta$$

The probability that a series of observations $(x_1, x_2, \ldots, x_m)$ leads to acceptance of H_0 is

$$P_{0,m} = \frac{1}{(2\pi)^{m/2} \sigma_0^m} \exp\left[-\frac{1}{2\sigma_0^2} \sum_{q=1}^{m} (x_q - \mu)^2 \right] \tag{3.99}$$

The probability of rejection of H_0 is

$$P_{1,m} = \frac{1}{(2\pi)^{m/2} \sigma_1^m} \exp\left[-\frac{1}{2\sigma_1^2} \sum_{q=1}^{m} (x_q - \mu)^2 \right] \tag{3.100}$$

Now

$$\ln \frac{P_{1,m}}{P_{0,m}} = m \ln \frac{\sigma_0}{\sigma_1} + \left(\frac{1}{2\sigma_0^2} - \frac{1}{2\sigma_1^2} \right) \sum_{q=1}^{m} (x_q - \mu)^2 \tag{3.101}$$

CI. The experiment is continued if

$$\ln B < \ln \frac{P_{1,m}}{P_{0,m}} < \ln A \tag{3.102}$$

$$\frac{2\ln B + m\ln(\sigma_1^2/\sigma_0^2)}{(1/\sigma_0^2)-(1/\sigma_1^2)} < \sum_{q=1}^{m}(x_q-\mu)^2 < \frac{2\ln A + m\ln(\sigma_1^2/\sigma_0^2)}{(1/\sigma_0^2)-(1/\sigma_1^2)} \tag{3.103}$$

or with

$$a_m = \frac{2\ln A + m\ln(\sigma_1^2/\sigma_0^2)}{(1/\sigma_0^2)-(1/\sigma_1^2)} \quad \text{and} \quad b_m = \frac{2\ln B + m\ln(\sigma_1^2/\sigma_0^2)}{(1/\sigma_0^2)-(1/\sigma_1^2)} \tag{3.104}$$

$$b_m < \sum_{q=1}^{m}(x_q-\mu)^2 < a_m \tag{3.105}$$

CII. The experiment is terminated and H_0 rejected if

$$\sum_{q=1}^{m}(x_q-\mu)^2 \geqslant a_m \tag{3.106}$$

TABLE 3.15.   Progress of a Sequential Analysis of Precision

$$\alpha = 0.05 \quad \beta = 0.10 \quad A = \frac{1-\beta}{\alpha} = \frac{0.90}{0.05} = 18 \quad B = \frac{\beta}{1-\alpha} = \frac{0.10}{0.95} = 0.105$$

$$\sigma_0 = 10 \quad \sigma_1 = 15 \quad \mu = 0 \quad L_0 = b_m \quad L_1 = a_m$$

$$L_0 = \frac{2\ln 0.105 + m\ln 2.25}{0.0055} = -811 + 146m$$

$$L_1 = \frac{2\ln 18 + m\ln 2.25}{0.0055} = 1040 + 146m$$

No. of observation m	Observation x_q	Sum $\sum_{q=1}^{m}(x_q-\mu)^2$	Decision	Conclusion	Result
1	7	49	CI	—	Continue
2	5	74	CI	—	Continue
3	8	138	CI	—	Continue
4	−11	259	CI	—	Continue
5	10	359	CI	—	Continue
6	8	423	CI	—	Continue
7	−9	504	CI	—	Continue
8	6	540	CI	—	Continue
9	−7	589	CI	—	Continue
10	6	625	CIII	H_0 accepted	Stop

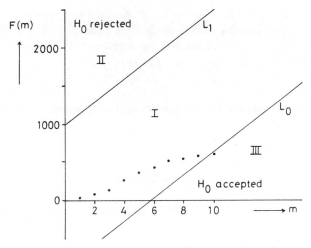

Figure 3.19. Sequential analysis of the precision of a series of observations.

CIII. The experiment is terminated and H_0 accepted if

$$\sum_{q=1}^{m} (x_q - \mu)^2 \leqslant b_m \qquad (3.107)$$

If the values of Table 3.15 are used, Figure 3.19 can be plotted.

The theory treated in the foregoing parts shows that it is possible to test both accuracy and precision of an analytical method in one experiment. However, the accuracy test is applicable only when the precision test has

TABLE 3.16. **Combined Sequential Analysis of Accuracy and Precision**

No. of Observation n, m	Observation	Decision	Conclusion	Result
1	7	AI, BI, CI	—	Continue
2	5	AI, BI, CI	—	Continue
3	8	AI, BIII, CI	—	Continue
4	−11	AI, CI	—	Continue
5	10	AI, CI	—	Continue
6	8	AI, CI	—	Continue
7	−9	AI, CI	—	Continue
8	6	AI, CI	—	Continue
9	−7	AIII, CI	—	Continue
10	6	CIII	$(H_0)_A, (H_0)_B$, and $(H_0)_C$ are true	Stop

shown that the precision of the tested method and standard method is equal. Also it is possible to prove that the precision of both methods is equal only when the accuracy test has shown that no bias exists. If the sequential analysis scheme of the preceding chapter is used, this means that a condition for acceptance of the hypothesis that both precision and accuracy of the tested and standard method are equal is that AIII and BIII and CIII decisions should be made. This is depicted in Table 3.16, a combination of Tables 3.14 and 3.15.

REFERENCES

AL 69 P. L. Alger: *Mathematics for Science and Engineering*, McGraw-Hill, New York, 1969.

AN 75 Analytical Chemistry: *Anal. Chem.*, **47**, 2527 (1975).

AN 76 Anal. Chem. Div. IUPAC: *Pure Appl. Chem.* **45**, 99 (1976).

BE 72 G. Bechtler: *Organisation et Automation des Laboratoires* (G. Siest, Ed.), Paris, 1972, p. 24.

BE 75 R. J. Bethea, B. S. Duran, T. L. Boullion: *Statistical Methods for Engineers and Scientists*, Marcel Dekker, New York, 1975.

BO 75 F. M. Bosch, J. A. Broekaert: *Anal. Chem.* **47**, 188 (1975).

CA 75 J. P. Cali et al.: *The Role of Standard Reference Materials in Measurement Systems*, N.B.S. Monograph 148, U.S. Govt. Printing Office, Washington, D.C., 1975.

CU 68 L. A. Currie: *Anal. Chem.* **40**, 586 (1968).

DA 71 O. L. Davies: *The Design and Analysis of Industrial Experiments*, Oliver and Boyd, Edinburgh, 1971.

DA 72 O. L. Davies: P. L. Goldsmith: *Statistical Methods in Research and Production*, Oliver and Boyd, Edinburgh, 1972.

DI 69 W. J. Dixon, F. J. Massey: *Introduction to Statistical Analysis*, 3rd ed., McGraw-Hill, New York, 1969.

DU 75 O. J. Dunn, V. A. Clark: *Applied Statistics: Analysis of Variance and Regression*, Wiley, New York, 1974.

DY 71 A. Dijkstra: "Analytische chemie en chemische analyse" (a survey in Dutch), *Chem. Weekbl.* A3–A5 (1971).

FE 69 R. W. Fennel, T. S. West: *Pure Appl. Chem.* **18**, 439 (1969).

GA 78 L. de Galan: *Chem. Weekbl.* m 447 (1978).

GI 63 P. M. E. M. van der Grinten: *Control Eng.* **10**(12), 51 (1963).

GR 63 P. M. E. M. van der Grinten: *Control Eng.* **10**(10), 87 (1963).

GR 65 P. M. E. M. van der Grinten: *J. Instr. Soc. Am.* **12**(1), 48 (1965).

GR 66 P. M. E. M. van der Grinten: *J. Instr. Soc. Am.* **13**(2), 58 (1966).

HA 76 R. Haeckel: *Rationalisierung des medizinischen Laboratoriums*, G-I-T. Verlag, Darmstadt, 1976.

HI 76 T. Hirschfeld: *Anal. Chem.* **48**(1), 16A (1976).

HO 70 T. Horne: *Z. Anal. Chem.* **252**, 241 (1970).

IN 74 J. D. Ingle, Jr.: *J. Chem. Ed.* **51**(2), 100 (1974).

KA 65 H. Kaiser: *Z. Anal. Chem.* **209**, 1 (1965).

KA 70 H. Kaiser: *Anal. Chem.* **42**(2), 24A (1970).

KA 73 H. Kaiser: *Methodicum Chimicum*, Vol. I, Analytical Part 1, Thieme-Verlag-Acad. Press, 1973, pp. 1–20.

KS 70 H. Kaiser: *Anal. Chem.* **42**(4), 26A (1970).

LE 71 F. A. Leemans: *Anal. Chem.* **43**(11), 36A (1971).

LI 73 C. Liteanu, I. Rică: *Mikrochim. Acta* **1973**, 745.

LI 75 C. Liteanu, I. Rică: *Mikrochim. Acta* **1975II**, 311.

MA 49 S. Marcuse: *Biometrics* **5**, 189 (1949).

MU 78 P. J. W. M. Müskens: *Anal. Chim. Acta* **103**, 445 (1978).

NA 63 M. G. Natrella: *Experimental Statistics*, National Bureau of Standards Handbook 91, U.S. Govt. Printing Office, Washington, D.C., 1963.

PA 72 D. A. Pantony, P. W. Harley: *Analyst* **97**, 497 (1972).

PL 46 R. L. Plackett, J. P. Burman: *Biometrika* **33**, 305 (1946).

PL 75 R. Plesch: *Glas-Instr. Tech.* **19**, 676 (1975).

PS 77 L. Pszonicki: *Talanta* **24**, 613 (1977).

PZ 77 L. Pszonicki, A. Lukszo-Bienkowska: *Talanta* **24**, 617 (1977).

SH 31 W. A. Shewhart: *Economic Control of the Quality of Manufactured Product*, Van Nostrand, New York, 1931.

SV 68 V. Svoboda, R. Gerbatsch: *Z. Anal. Chem.* **242**, 1 (1968).

VA 77 B. G. M. Vandeginste: *Anal. Lett.* **10**(9), 661 (1977).

WA 47 A. Wald: *Sequential Analysis*, Wiley, New York, 1947.

YO 51 W. J. Youden: *Statistical Methods for Chemists*, Chapman and Hall, London, 1951.

YO 72 W. J. Youden: *Statistical Techniques for Collaborative Tests*, A.O.A.C., Washington, D.C., 1972.

YO 75 W. J. Youden, E. H. Steiner: *Statistical Manual of the Association of Official Analytical Chemists*, A.O.A.C., Washington, D.C., 1975.

DATA PROCESSING

The first task of an analyst who wants to analyze a process or procedure is to collect relevant data of measurements on that process or procedure. The questions what data are relevant and how many measurements are required must be assumed from consideration of information theory and the sampling strategy (see Chapter 2 and Section 4.2). The reduction of the data collected according to a certain sampling strategy without loss of relevant information is the subject of this chapter. However, before the discussion of data reduction methods we must consider the methods of data production.

4.1 DATA PRODUCTION

There are various methods of data production:

1. A measurement is made and the outcome of the measurement is found from the position of a dial on a scale of an instrument. It is the analyst who estimates the position of the dial and converts this to numbers in a digital notation.

2. The measurement is made continuously and translated in the position of a recorder pen on a strip of recorder paper. The analyst reads the position of the recorded pattern on the time base at times of interest and converts the selected pen positions to digital numbers.

3. The measurement is made with a fixed frequency, converted electronically into a digital notation, and recorded on a device such as cassette tape, disk, paper tape, or magnetic tape. The analyst is not involved in this process, apart from initiating the procedure of data collection.

The second task, after the numbers are obtained, is to reduce the number of measurements, remove the erroneous and irrelevant data, and convert the measurements into statements pertaining to the condition of the process or procedure under control. In modern analytical procedures, a data acquisition rate of 10^5 measurements per second is not unusual. In order to convert this enormous amount of data, collected, for example, during one day, into a form that can be handled, data reduction procedures should be applied. Depending on the aim of the analysis, various data reduction procedures may be applied.

4.1.1 Tests for Normality

Since many statistical tests can be applied only when the set of data is normally (Gaussian) distributed, it is recommended that a test for normality be made first. Nowadays computations of statistical parameters are often performed routinely by application of standard calculator or computer subroutines. Nowhere in the instruction manuals is it adequately stressed that the application of these methods presupposes normal distribution.

In the literature a number of methods to test for normality has been described. An oft-used test is that for skewness. It signals asymmetry of the distribution. Although a passed test for skewness doesn't necessarily prove that the distribution is normal, a failed test obviously indicates that normal statistics should be applied with care.

A parameter for skewness is the so-called "third moment," denoted by μ_3:

$$\mu_3 = \frac{\sum_{i=1}^{n} (x_i - \mu)^3}{n} \tag{4.1}$$

An index of skewness that is independent of the unit of measurement is the third moment divided by the cube of the standard deviation. This parameter is denoted by $\sqrt{\beta_1}$:

$$\sqrt{\beta_1} = \left(\frac{\mu_3}{\sigma^3}\right)^{1/2} \tag{4.2}$$

The value of $\sqrt{\beta_1}$ should be zero, but since σ normally is not known and is estimated by s, and because the third moment is also an estimation, some deviation from zero is allowed. Table 4.1 gives the probability of various deviations from zero for a normal distribution. Other, more reliable tests on normality are summarized in Table 4.2 (AM 77). As seen from this table, all methods are based on cumulative distribution functions, except the Shapiro–Wilk method. (Cumulative frequencies are obtained by adding for each class all the class frequencies up to that point, divided by

TABLE 4.1. $\sqrt{\beta_1}$ Values for Different Levels of Significance α

n	$\alpha = .95$	$\alpha = .99$
25	0.711	1.061
30	0.661	0.982
35	0.621	0.921
40	0.587	0.869
45	0.558	0.825
50	0.533	0.787
60	0.492	0.723
70	0.459	0.673
80	0.432	0.631
90	0.409	0.596
100	0.389	0.567
150	0.321	0.464
200	0.280	0.403
300	0.230	0.329
400	0.200	0.285
500	0.179	0.255
1000	0.127	0.180

Because the distribution of $\sqrt{\beta_1}$ is symmetrical about zero, the same values, with negative sign, correspond to the lower limits.

the total number of data in the data set.) In the literature surveyed, there is no overall agreement as to which method is "best" for all possible empirical distributions (DI 69, DY 74, SH 72). Comparison of the various normality tests results in the observation that the data set is considered to be normal by some methods and not normal by others. However, the Chi-square test is generally regarded to be inferior to all other tests.

Here only the Kolmogorov test is considered in more detail because it is generally applicable, easy to use, and one of the dependable ones.

Graphic Kolmogorov Test

1.   Arrange the data in ascending order.

2.   Draw the cumulative frequency line for a normal distribution (G in Figure 4.1).

3.   Draw two lines parallel to the line of step 2, each at a distance D. D can be obtained from Table 4.3.

4.   Draw the cumulative frequency line of the ordered data (S in Figure 4.1).

5.   If S is found somewhere outside the lines of step 3, the distribution can be considered not normal with a probability given by the chosen value of Table 4.3).

TABLE 4.2. Principal Methods for Normality Testing (AM 77)

Test Method	Computing Formula	Symbols Used	Remarks	Refs.
Chi-square	$\chi^2 = \sum \dfrac{(O_j - T_j)^2}{T_j}$	O_j = Observed class frequency T_j = Theoretical class frequency	The data are arranged in descending order and then subdivided into arbitrary classes, each containing at least 5 numbers. The "observed frequencies" are the number of data in each class. The "theoretical frequencies" are those based on the cumulative normal distribution function. The computed χ^2 is evaluated using a χ^2 table at $k-3$ degrees of freedom, where k = number of classes. This method requires a relatively large sample size (at least 15–30) and is considered a test of low power	DI 69, LI 67, CO 71
Lilliefors	$\max D_i^- = S(x_i) - \dfrac{i-1}{n}$	i = sequential data points or frequencies $(i-1)/n$ = cumulative distribution values of the data set $S(x_i)$ = corresponding normal cumulative distribution values	The data are arranged in ascending order and the D_j^--values computed by the formula given, i.e., not considering the first data cell. The largest D^--value obtained is evaluated using the Lilliefors table. This method can be used with data sets given in terms of frequencies or individual data. Sample sizes as low as 4 can be tested. This method will fail if the first data cell contains a disproportionately large number of data compared to the rest of the set	LI 67, CO 71

137

TABLE 4.2. (continued)

Test Method	Computing Formula	Symbols Used	Remarks	Refs.
Kolmogorov D^+ (Kolmogorov one-sample statistics)	$\max D_i^+ = S(x_i) - \dfrac{i}{n}$	Symbols as given, and $i/n =$ cumulative distribution values of the data set	The data are arranged in ascending order and the D_i^+-values computed, as given by the formula. The largest D_i^+-value obtained is evaluated using the Kolmogorov table (two-sided). The method is usable with sample sizes as low as 2, with data given individually or as frequency distributions. This method can supplement the Lilliefors test if it is suspected that it may fail. A modified version of this test multiplies the largest D_i^+-value by a correction factor and evaluates this modified value using a special table	ST 70, ST 74, CO 71, MA 51, OW 62, SI 56
Kolmogorov D (modified Kolmogorov, E_n)	$D_i = \max(D^+, D^-)$	Symbols as given	The largest D_i^+ or D_i^-, calculated by the above formulas, is chosen. This value is multiplied by a correction factor and evaluated using a special table. Usable for individual data or those given as frequency distributions	ST 70, ST 74
Kuiper	$V = D^+ + D^-$	Symbols as given	The sum of the largest D_i^+ and D_i^- is calculated by the formulas given. This sum is multiplied by a correction factor and evaluated using a special table. Usable for individual data or those given as frequency distributions	ST 70, ST 74

138

Cramér–von Mises (Cramér–Smirnov)	$$W^2 = \frac{1}{12_n} + \sum\left(S(x_i) - \frac{i-1}{n}\right)^2$$	Symbols as given, and $(i-1)/n$ = cumulative distribution values of the data set	The data are arranged in ascending order. The W^2 computed is evaluated using appropriate tables. From this W^2, a modified W^2 can be calculated, and this is evaluated using a special table. With the standard table, sample sizes as low as 2 can be evaluated. Usable for individual data or those given as frequency distributions	ST 68, ST 70, ST 74, CO 71, MA 51, OW 62, SI 56
Watson	$$U^2 = W^2 - n\left(\bar{S}(x) - \tfrac{1}{2}\right)^2$$	Symbols as given, and $\bar{S}(x) = S(x_i)/n$	This is a modification of the Cramér–von Mises method. The computed U^2 is modified and this modified U^2 is then evaluated using a special table. Usable for individual data or those given as frequency distribution	ST 70, ST 74
Anderson–Darling	$$A^2 = -\left\{\sum(2_i - 1)[\ln S(x_i) + \ln[1 - S(n+1-i)]]\right\}\frac{}{n} - n$$	Symbols as given	The data are arranged in ascending order. The A^2 computed is modified and this modified value is evaluated using a special table. Usable for individual data or those given as frequency distributions	ST 70, ST 74
Shapiro–Wilk	For even number of data: $n = 2K$ $$W = \frac{\left[\sum_1^k a_{n-i+1}(x_{n-i+1} - x_i)\right]^2}{(x_i - \bar{x})^2}$$ For odd number of data: $n = 2K + 1$ $$W = \frac{[a_n(x_n - x_1) + \cdots + a_{k+2}(x_{k+2} - x_k)]^2}{\sum(x - \bar{x})^2}$$	x_i = data $\bar{x} = x_i/n$ a_n = special coefficients	The data are arranged in ascending order. The special coefficients needed for the calculation and the table needed to evaluate the W computed are accessible only via the references listed. Samples as low as 3 can be evaluated	SH 65, SH 68, AG 71

NOTE: The various tests for normality have been reviewed recently by Shapiro et al. (SH 65, SH 68, AG 71), Schuster (SC 73), Dyer (DY 74), Shapiro and Francis (SH 72), Klimko and Antle (KL 75), and Govindarajulu (GO 76).

Chemometrics: Theory and Application

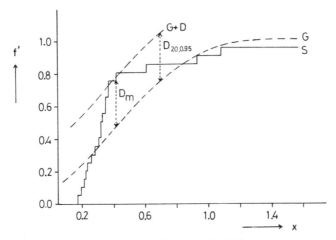

Figure 4.1. Cumulative frequency for Kolmogorov test.

Numerical Kolmogorov Test

1. Arrange the data in ascending order.
2. Compute $D_i = S(x_i) - i/n$, with i = sequential data points or frequencies, n = total number of data points, and $S(x_i)$ = corresponding normal cumulative distribution values.
3. Select the maximum value (D_m) of D_i.

TABLE 4.3. D Values for Different Levels of Significance α

n	$\alpha = .90$	$\alpha = .95$	$\alpha = .99$
1	0.950	0.975	0.995
2	0.776	0.842	0.929
3	0.642	0.708	0.828
4	0.564	0.624	0.733
5	0.510	0.565	0.669
6	0.470	0.521	0.618
7	0.438	0.486	0.577
8	0.411	0.457	0.543
9	0.388	0.432	0.514
10	0.368	0.410	0.490
12	0.338	0.375	0.450
14	0.314	0.349	0.418
16	0.295	0.328	0.392
18	0.278	0.309	0.371
20	0.264	0.294	0.356
>20	$1.22/\sqrt{n}$	$1.36/\sqrt{n}$	$1.63/\sqrt{n}$

TABLE 4.4. Numerical Kolmogorov Test[a]

x_i (ordered)	i	i/n	$S(x_i)$	$|D_i|$
0.17	1	0.05	0.221	0.171
0.19	2	0.10	0.239	0.139
0.21	3	0.15	0.255	0.105
0.22	4	0.20	0.264	0.064
0.23	5	0.25	0.274	0.024
0.26	6	0.30	0.302	0.002
0.28	7	0.35	0.323	0.027
0.31	8	0.40	0.352	0.048
0.32	9	0.45	0.363	0.087
0.32	10	0.50	0.363	0.137
0.33	11	0.55	0.374	0.176
0.35	12	0.60	0.397	0.203
0.35	13	0.65	0.397	0.253
0.37	14	0.70	0.417	0.283
0.37	15	0.75	0.417	0.333*
0.42	16	0.80	0.476	0.324
0.61	17	0.85	0.677	0.173
0.93	18	0.90	0.915	0.015
1.08	19	0.95	0.963	0.013
1.57	20	1.00	0.999	0.001

[a] $m = 0.445$, $\sqrt{\beta_1} = 1.93$, $s = 0.354$, $D_m = 0.333$.

4. If D_m is larger than the value given by Table 4.3, the distribution can be considered as not normal, with a probability given by the selected α-value of Table 4.3.

An example of both methods is given in Figure 4.1 and Table 4.4.

4.1.2 Transformation of Data

Occasionally one encounters a distribution that, by one of the tests mentioned above, departs too far from normality. It would not be safe then to apply the common statistical tests, and it would be a great advantage if, by a simple transformation of the variable, an approximately normal distribution could be obtained. For example, if the weights of a collection of spheres were given and found not to be normally distributed, a natural transformation would be to take the cube root of the weight. Here one tests for the diameters to be normally distributed. Often a skewness may be removed or reduced by using the logarithm of the observation. The new distribution will not necessarily be normal, but will probably pass the tests for normality. Some other typical transformations to normalize a skew distribution use the square, the square root, or the reciprocal of the variable.

Box and Cox (BO 64, SL 73) described an universally applicable transformation for nonnormal data sets to create sets that are as near to normality as possible. For theoretical reasons they prefer to write:

$$y_i^{(\lambda)} = \frac{(x_i + c)^\lambda - 1}{\lambda} \qquad \text{for } \lambda \neq 0 \qquad\qquad (4.3)$$

$$y_i^{(\lambda)} = \log(x_i + c) \qquad \text{for } \lambda = 0 \qquad\qquad (4.4)$$

with x_i = the measured quantity
$\quad y_i^{(\lambda)}$ = the transformed datum
$\qquad \lambda$ = the transformation parameter
$\qquad c$ = constant
For practical applications this can be reduced to

$$y_i^{(\lambda)} = (x_i)^\lambda \qquad \text{for } \lambda \neq 0 \qquad\qquad (4.5)$$

$$y_i^{(\lambda)} = \log x_i \qquad \text{for } \lambda = 0 \qquad\qquad (4.6)$$

The transformation is applied for a range of values of λ and each time a test for normality is performed. The best value of λ is chosen.

Box and Cox apply a somewhat different criterion in situations where on the data an analysis of variance should be performed. Note that the transformation rules mentioned above are implied in the often used "intuitive" transformations:

λ	Transformation
2	Square
1	Unaltered
0.5	Square root
0.333	Cube root
0	Log
−1	Reciprocal

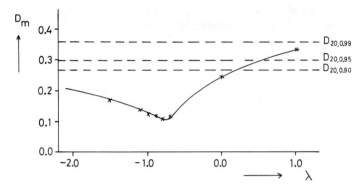

Figure 4.2. Estimation of optimal transformation factor.

In Figure 4.2 an example is given of data transformation and the various test results obtained. The original data are those of Table 4.4. As shown in Figure 4.2, the values of D_m of the Kolmogorov test have a minimum at $\lambda = -0.8$. This means that a transformation according to $y_i^{(\lambda)} = (x_i)^{-0.8}$ gives a distribution that, according to the Kolmogorov test, is as near as possible to normal. See Figure 4.3 and Table 4.5. However, the values of $D_{n,\alpha}$ are such that, for example, for $D_{20,0.95}$ all distributions obtained by values of λ between $-2.5 < \lambda < 0.5$ should be considered as normal. Thus an "intuitive" transformation by taking the square root ($\lambda = 0.5$), the logarithm ($\lambda = 0$), or the reciprocal ($\lambda = -1$) would suffice as well.

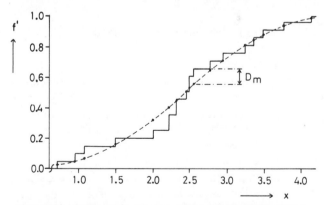

Figure 4.3. Kolmogorov test on transformed data of Figure 4.1.

TABLE 4.5. Transformation of Data to a Normally Distributed Data Set

$\lambda = 1$ x_i	$\lambda = -0.8$ y_i^λ
0.17	4.127
0.19	3.776
0.21	3.485
0.22	3.358
0.23	3.241
0.26	2.938
0.28	2.769
0.31	2.552
0.32	2.488
0.32	2.488
0.33	2.428
0.35	2.316
0.35	2.316
0.37	2.215
0.37	2.215
0.42	2.002
0.61	1.485
0.93	1.060
1.08	0.940
1.57	0.697
$m = 0.445$	$m = 2.445$
$s = 0.354$	$s = 0.918$
$\sqrt{\beta_1} = 1.93$	$\sqrt{\beta_1} = -0.177$
$D_m = 0.333$	$D_m = 0.102$

4.2 INFORMATION THEORY

According to a definition of Gottschalk (GO 72) "analytical chemistry provides us with optimal strategies for collecting and appreciating relevant information on situations and processes in materials." The input into the analytical procedure contains some actual information and some latent information (uncertainty). During the analytical process the actual information increases at the expense of the latent information. For example, a sample is analyzed for iron content. The actual information here is that iron is expected to be present and the amount is to be determined. The latent information is the unknown quantity of iron. By an analytical procedure the iron content is measured and found to be between $10.82\pm$

0.01% weight. Thus the uncertainty is decreased from 0–100% to 10.81–10.83% at the expense of the latent information.

Because the output is information that is disclosed by measurements the definition of this quantity needs a closer examination. One definition stems from information theory (SH 49, PE 67): "Information equals diminishing uncertainty." In our example the iron content before analysis equals 0–100% and afterward 10.81–10.83%. Using a more general definition information I equals uncertainty $H_{before} - H_{after}$ analysis. Another name for uncertainty H is entropy. The magnitude of H is given by the number of possibilities (n) before and after the experiment, respectively. This implies H and also I to be functions of n

$$I = F(n)$$

$$I = F(1) = 0$$

Only one possibility offers certainty!

In order to find the function $F(n)$ two independent experiments are considered, one selecting a single possibility from a collection of n_1, and the other selecting a single possibility from a collection of n_2; thus

$$I_1 = F(n_1)$$

$$I_2 = F(n_2)$$

A combination of these two experiments is represented by a single selection from $n = n_1 \times n_2$ possibilities. According to our definition

$$I_{tot} = I_1 + I_2$$

However,

$$F(n) = F(n_1 * n_2) = F(n_1) F(n_2)$$

This implies I and H should be logarithmic functions in n. It is assumed that $I = k \log_z n$, where k and z are constants to be defined.

The most simple experiment is a decision between two alternatives yes or no. The yield of information of such an experiment is selected as a unit of information, the bit (binary unit) (EC 71–EC 73, EK 72). This selection implies

$$I = k \log_z 2 = 1 \text{ (bit)} \tag{4.7}$$

and $k = 1/\log_z 2$. Thus

$$I = \frac{\log_z n}{\log_z 2} = \log_2 n = \text{ld } n \tag{4.8}$$

Here ld represents the logarithm with base 2. Other units found in the literature are "nit," based on logarithm with base e, and "dit," based on information obtained by selecting 1 out of 10 independent possibilities.

If a selection must be made of one out of n possibilities with unequal probabilities, the equation of Shannon (SH 49) should be applied:

$$I = H = - \sum_{i=1}^{n} P_i \, \mathrm{ld} \, P_i \qquad (4.9)$$

This is illustrated in the following example: a selection is made of one out of three possibilities (A, B, and C). The possibility of occurrence of possibility A equals P_A, of B P_B, and of C P_C, $P_A \neq P_B \neq P_C$. Performing the experiment n times yields n_1 times A so $P_A = n_1/n$. The same holds for n_2 and P_B and n_3 and P_C. This means

$$I_A = \mathrm{ld} \frac{n}{n_1}$$

$$I_B = \mathrm{ld} \frac{n}{n_2}$$

and

$$I_C = \mathrm{ld} \frac{n}{n_3} \qquad (4.10)$$

The average value of the information gain is found by summation of the individual informations, multiplied by a factor that accounts for the frequency of occurrence. Thus

$$I_{\mathrm{tot}} = - \frac{n_1}{n} \mathrm{ld} \frac{n_1}{n} - \frac{n_2}{n} \mathrm{ld} \frac{n_2}{n} - \frac{n_3}{n} \mathrm{ld} \frac{n_3}{n}$$

$$= - \sum_i P_i \, \mathrm{ld} \, P_i \qquad (i = A, B, C) \qquad (4.11)$$

This equation applied to two situations with unequal probability yields

$$H = - \sum_{i=1}^{n} P_i \, \mathrm{ld} \, P_i \qquad (4.12)$$

$$= - \left[P_i \, \mathrm{ld} \, P_i + (1 - P_i) \mathrm{ld} (1 - P_i) \right]$$

Figure 4.4 demonstrates that the entropy (and corresponding information) gets a maximal value of 1 for equal possibilities ($P = \frac{1}{2}$)

$$H = 2 \left(- \frac{1}{2} \mathrm{ld} \frac{1}{2} \right) = 1 \text{ (bit)}$$

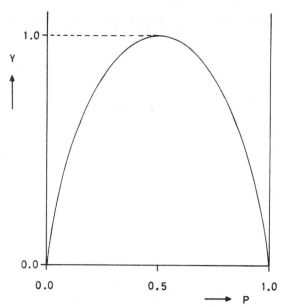

Figure 4.4. Entropy y as a function of probability of occurrence P of one out of two situations.

Another possibility to be mentioned is a probability distribution represented by a Gaussian function. Such a distribution may be approximated by means of a histogram with a given width of the classes Δx. The smaller Δx, the more classes needed and the better the approximation. The probability of occurrence of a given class equals

$$\rho_i = \int_{x_i}^{x_c + \Delta x} \rho(x_i)\,dx = \rho(x_i)\Delta x \tag{4.13}$$

The entropy of a real continuous distribution may be approximated by the entropy of the histogram

$$H = -\sum_i P_i \operatorname{ld} P_i = -\sum_i [P(x_i)\Delta x]\operatorname{ld}[P(x_i)\Delta x] \tag{4.14}$$

which for small values of Δx becomes

$$H = -\int_{-\infty}^{+\infty} P(x)\operatorname{ld}[P(x)\Delta x]\,dx$$

$$= -\int_{-\infty}^{+\infty} P(x)[\operatorname{ld} P(x)]\,dx - \int_{-\infty}^{+\infty} P(x)[\operatorname{ld}\Delta x]\,dx \tag{4.15}$$

a constant x-value over the entire integration range implies the integrals

$$H = - \int_{-\infty}^{+\infty} P(x)[\mathrm{ld}\, P(x)]\, dx - \mathrm{ld}\, \Delta x \qquad (4.16)$$

because of $\int_{-\infty}^{+\infty} P(x)\, dx = 1$.

It is seen that a limiting value of $\Delta x \to 0$ results in a growth toward infinity of $\mathrm{ld}\, \Delta x$, which is not realistic. It means that the value of H can be known apart from the value of $-\mathrm{ld}\, \Delta x$. The value of this term depends on the Δx-value chosen and is called "zero-level standard" entropy. Its actual value may be considerable! The applicability of this equation is not hampered by the unknown term because as a rule one is interested in differences of information before and after analysis. In subtracting two informations the constant disappears.

4.2.1   Application of Information Theory to Quantitative Analysis

The constant P_c of a certain compound in a mixture before and after analysis is depicted in Figure 4.5. Before the analysis each concentration c_0 and c_e has an equal probability of occurrence; after analysis the concentration is known with a deviation of δ_c.

Application of the Shannon equation on the situation before analysis yields

$$
\begin{aligned}
H_{\text{before}} &= - \int_{-\infty}^{+\infty} P_c(\mathrm{ld}\, P_c)\, dc - \mathrm{ld}\, \Delta x \\
&= - \int_{-\infty}^{c_0} P_c(\mathrm{ld}\, P_c)\, dc - \int_{c_0}^{c_e} P_c(\mathrm{ld}\, P_c)\, dc - \int_{c_e}^{+\infty} P_c(\mathrm{ld}\, P_c)\, dc - \mathrm{ld}\, \Delta x \\
&= 0 - P_b(c_e - c_0)\, \mathrm{ld}\, P_b - 0 - \mathrm{ld}\, \Delta x \qquad (4.17)
\end{aligned}
$$

and after analysis

$$H_{\text{after}} = 0 - P_e(\mathrm{ld}\, P_e)\delta_c - \mathrm{ld}\, \Delta x \qquad (4.18)$$

Before analysis

$$\int_{c_0}^{c_b} P_b\, dc = 1$$

or

$$P_b = \frac{1}{c_e - c_0} \qquad (4.19)$$

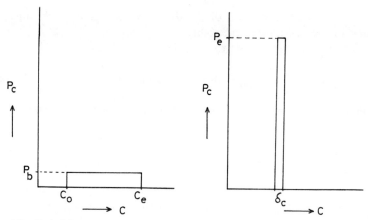

Figure 4.5. Probability of occurrence of a concentration P as a function of the numerical value of that concentration C before (P_b) and after (P_e) analysis.

which on substitution in eq. 4.17 yields

$$H_{\text{before}} = \frac{1}{c_e - c_0} \left[\text{ld}(c_e - c_0) \right] (c_e - c_0) - \text{ld}\, \Delta x$$

$$= \text{ld}(c_e - c_0) - \text{ld}\, \Delta x \tag{4.20}$$

After analysis

$$P_e \delta_c = 1$$

and thus

$$H_{\text{after}} = \text{ld}\, \delta_c - \text{ld}\, \Delta x \tag{4.21}$$

The information gain I becomes

$$I = H_{\text{before}} - H_{\text{after}}$$

$$= \text{ld}\, \frac{c_e - c_0}{\delta_c} \tag{4.22}$$

Application of this procedure in a situation giving a Gaussian distribution of the concentration after analysis implies

$$P_c = \frac{1}{\sigma \sqrt{2\pi}} \exp\left(-\frac{(c - \mu)^2}{2\sigma^2} \right) \tag{4.23}$$

in which 2σ equals the width of the distribution at half height and μ the position of the maximum probability.

This distribution results in an H_a-value

$$H_a = \text{ld}(\sigma\sqrt{2\pi e}) - \text{ld}\,\Delta x \tag{4.24}$$

and thus

$$I = \text{ld}\left(\frac{c_e - c_0}{\sigma\sqrt{2\pi e}}\right) \tag{4.25}$$

In situations where σ is unknown, its value may be estimated from n_s measurements and the mean result of the measurements $\bar{c}$:

$$s = \left[\frac{\sum_{i=1}^{n_s}(c_i - \bar{c})^2}{n_s - 1}\right]^{1/2} \tag{4.26}$$

A somewhat more general expression is given by

$$I = \text{ld}\frac{(c_e - c_0)\sqrt{n_s}}{2t_{p,\nu}s} \tag{4.27}$$

Here $t_{p,\nu}$ represents the probability distribution of the analytical results p and a parameter ν, representing the number of degrees of freedom.

Thus the gain of information depends on s and on n. It is seen (eq. 4.27) that an increase in the number of measurements is not proportional to the increase of information: the increase of information is less than the information expected from the number of measurements. The diminishing effect of the estimated number of measurement on information gain is called redundancy R:

$$R = I_{\text{max}} - I_n$$

$$= nI - I_n$$

This equation implies that the maximum amount of information to be gained from n measurements equals n times the information of one single measurement. The interaction between successive measurements thus is supposed to be absent. Instead of the redundancy R sometimes the relative redundancy ρ is used (KA 70, KI 70):

$$\rho = \frac{I_{max} - I_n}{I_{max}} = 1 - \frac{I_n}{nI_1} \qquad (4.28)$$

The redundancy provides a method for estimation of the mutual interaction of successive measurements. A high value of the redundancy implies that the effects of interruptions otherwise leading to erroneous results are partially cancelled by the correct results in a series of measurements. In quantitative analysis a calibration graph frequently is used. The question arises, how is the information theory applied here? To begin, the standardized function $u = ac + b$ is constructed from m samples with concentrations c_i and measurements u_i. The analysis of a sample yields a measurement u_a and thus a concentration c_a. A set of n parallel measurements result in $\bar{u}_a$ and $\bar{c}_a$. With a linear dependence of c on u the error in c_a, Δc_a, equals

$$\Delta c_a = \frac{t_{p,\nu} s_u}{a} \left\{ \frac{1}{n} + \frac{1}{m} + \frac{(\bar{u}_a - \bar{u})^2}{a^2 \left[\sum\limits_{i=1}^{m} c_i^2 - \left(\sum\limits_{i=1}^{m} c_i \right)^2 / m \right]} \right\}^{1/2} \qquad (4.29)$$

The actual concentration equals $c_a \pm \Delta c_a$. Substitution into eq. 4.29 with $s_u t_{p,\nu} / \sqrt{n} = \Delta c_a$ results in

$$I = \mathrm{ld} \left[\frac{(c_1 - c_0)a}{2 s_u t_{p,\nu}} \right] - \tfrac{1}{2} \mathrm{ld} \left[\frac{1}{n} + \frac{1}{m} + \frac{(\bar{u}_a - \bar{u})^2}{a^2 \left[\sum\limits_{i}^{m} c_i^2 - \left(\sum\limits_{i}^{m} c_i \right)^2 / m \right]} \right] \qquad (4.30)$$

This implies that the information increases upon

- A higher value of the sensitivity a of the analysis.
- An increase of n, the number of duplicate measurements.
- A decrease of s_u, which means an increase in reliability.
- A higher number of testing samples m.
- A higher range of combinations of the testing samples.
- Measuring close to the center of gravity of the test set $(\bar{u}_a - \bar{u})$ becomes smaller.

A small difference between c_e and c_0 results in a decrease in information. This disadvantage may partly be compensated by centering the testing samples around the expected value of the analysis. The most important observation to be made from eq. 4.29 is the overriding effect of sensitivity a and reliability s_u in comparison with the number of test samples m or number of duplicates n. Based on eq. 4.29 an optimal value of m, the number of test samples, can be derived for a given value of a, $c_e - c_0$, $\bar{u}_a - u$, n, s_u, and $t_{p,v}$ and a required information in, for example, $I \geqslant 9.5$ (Table 4.6).

As an example, one fake coin is to be selected out of a collection of 39, without knowing the deviation of weight. The selection method, weighing by an analytical balance, offers three different possibilities: light, heavy, and equal weights. The amount of information required for solving the problem is the reduction of 78 possibilities to one—each coin may be too light or too heavy—$I = \mathrm{ld}\,78 - \mathrm{ld}\,1 = 6.285$. This information is gained by four weighings, because 3^4 equals 81 possibilities. In order to use the available information as well as possible, the first weighing should exclude at least $78 - 3^3 = 51$ possibilities. The information gain thus should be $I_1 = \mathrm{ld}\,78 - \mathrm{ld}\,27 = 1.5305$. This is achieved by balancing two groups of 13 coins; the third group is put aside. Now there are two possibilities:

1. Upon balancing the coins, one decides that the fake coin should be found from the group of 13 coins put aside. This implies 26 possibilities and $I_{1_1} = \mathrm{ld}\,3 = 1.585$ (bit).

TABLE 4.6.   Derivation of the Optimal Number of Test Samples[a]

$c_e - c_0 = 100$ α in $t_{p,v} = 0.039$
$\bar{u}_a - u = 0$
$s_u = 0.01$
$a = 0.5$

I (bit)	$n=1$	$n=2$	$n=3$	$n=4$	$n=5$	$n=10$	$n=100$	$n=\infty$
$m=3$	7.04	7.40	7.54	7.63	7.70	7.85	8.02	8.04
$m=4$	8.83	9.20	9.38	9.49	9.56	9.75	9.96	9.99
$m=5$	9.33	9.72	9.91	10.04	10.12	10.32	10.59	10.62
$m=6$	9.60	9.99	10.21	10.34	10.43	10.66	10.96	11.00
$m=7$	9.73	10.14	10.35	10.50	10.60	10.84	11.18	11.23
$m=10$	9.92	10.36	10.60	10.75	10.80	11.15	11.59	11.65
$m=100$	10.19	10.68	10.99	11.17	11.32	11.79	13.01	13.65
$m \to \infty$[b]	10.24	10.74	11.03	11.24	11.40	11.90	13.56	—

[a]When a straight line calibration curve is employed the number of degrees of freedom v equals $(m-2)$.
[b]$m \to \infty$ holds when the calibration coefficient is known exactly, for example, because of a theoretical constant.

2. Nonbalance produces three groups of 13 coins; the group put aside is formed by 13 correct coins. This excludes 26 possibilities. The two remaining groups are correct or too heavy and correct or too light which diminishes the uncertainty to 26 possibilities. Thus $I_{1_2} = I_{1_1} = 1.585$ (bit).

The three weighings left for solving the problem offer $3^3 = 27$ possibilities. The second weighing should exclude at least $26 - 3^2 = 17$ possibilities, which means an information gain $I_2 = \mathrm{ld}\,26 - \mathrm{ld}\,9 = 1.5305$. Depending on the result of these two weighings 17 or 18 additional possibilities are excluded, resulting in eight or nine possible situations remaining. The possible situations range from four coins of which nothing is known except that the group contains the fake coin to nine coins of which four are known to be possibly too heavy and five to be possibly too light (or vice versa).

The third solution, depending on the history, selects three groups of three coins each or one group of three and one group of one. The number of possibilities is reduced to two or three, and the last weighing selects the faked coin and denotes the sign of its deviating weight.

The selection method is summarized in Table 4.7.

4.2.2 Application of Information Theory to Qualitative Analysis

The measurement of the identity of a given sample is a qualitative analysis. Cleij and Dijkstra (CL 79) treated the theory on the application of information theory. Here only some practical examples are given. In order to fix the identity of a compound, information is required, for example, the measurement of the melting point. The amount of information thus obtained depends on the probability of occurrence of that particular melting point.

In practice melting points cannot be measured exactly; therefore the range over which they are distributed is subdivided into a number of classes. The number of compounds with a melting point located in a given class gives the probability of occurrences of a particular melting point. To illustrate this a histogram has been constructed starting with 1349 arbitrarily chosen compounds. From this set the melting points of 348 compounds are unknown. The melting points of the other compounds are distributed over 100 classes with a class width of $5°C$ ranging from $-150°C$ up to $+350°C$ (see Figure 4.6).

Assuming an accuracy of the m.p. measurement corresponding with the width of the classes in the histogram, the information yield of a m.p. measurement equals

$$I = -\sum p_i \mathrm{ld}\, p_i$$

$$= 6.17 \text{ (bit)}$$

A more accurate measurement of the melting point results in a relatively smaller reliability interval in comparison with the width of a class and correspondingly

TABLE 4.7. Weighing Scheme [a]

Start
13 vs. 13;

$$I_0 = \log_2 78 = 6.285 \text{ (bit)}$$

13

Balance → $I_1 = \mathrm{ld}\,78/26 = \mathrm{ld}\,3$ ($13\uparrow + 26\varnothing$)

Nonbalance → $I_1 = \mathrm{ld}\,78/26 = \mathrm{ld}\,3$ ($13\uparrow + 13\downarrow + 13\varnothing$)

$4? + 1$ vs. $5?$; $4? + 25\varnothing$

$5\uparrow + 4\downarrow$ vs. $5\uparrow + 4\downarrow$; $3\uparrow + 5\downarrow + 13\varnothing$

Balance
$I_2 = \mathrm{ld}\,26/8$
($4? + 35\varnothing$)

Nonbalance
$I_2 = \mathrm{ld}\,26/9$
($4\uparrow + 5\downarrow + 30\varnothing$)

Balance
$I_2 = \mathrm{ld}\,26/8$
($3\uparrow + 5\downarrow + 31\varnothing$)

Nonbalance
$I_2 = \mathrm{ld}?26/9$
($5\uparrow + 4\downarrow + 30\varnothing$)

$1\uparrow + 2\downarrow$ vs. $2\downarrow + 1\uparrow$; $1\uparrow + 1\downarrow + 31\varnothing$ (or $2\uparrow + 1\downarrow + 30\varnothing$ or $3\uparrow + 30\varnothing$)

Nonbalance
$(2\uparrow + 1\downarrow + 36\varnothing)$

Balance
$I_3 = \mathrm{ld}\,I_i = \mathrm{ld}\,\dfrac{78}{26} * \dfrac{8}{2} * 2 = \mathrm{ld}\,78$
$(1\uparrow + 1\downarrow)$

$3?$ vs. 3; $1? + 32\varnothing$

Balance
$I_3 = \mathrm{ld}\,8/2 = \mathrm{ld}\,4$
($1? + 38\varnothing$)

Nonbalance
$I_3 = \mathrm{ld}\,8/3$
($3\uparrow + 36\varnothing$)
(or $2\uparrow + 1\downarrow + 36\varnothing$)
or $3\uparrow + 36\varnothing$)

1 vs. 1?; 38 $\varnothing$

$1\uparrow$ vs. $1?$; $1\uparrow + 36\varnothing$
or $1\downarrow + 36\varnothing$
$I_4 = \mathrm{ld}\,3$

$I_4 = \mathrm{ld}\,2 = 1$

Total gain $I = \displaystyle\sum_{i=1}^{4} I_i = \mathrm{ld}\,\dfrac{78}{26} * \dfrac{26}{8} * \dfrac{8}{2} * 2 = \mathrm{ld}\,78$

or $\mathrm{ld}\,\dfrac{78}{26} * \dfrac{26}{9} * \dfrac{9}{3} * 3$

[a] $\varnothing$ = decided, ? = not yet decided.

154

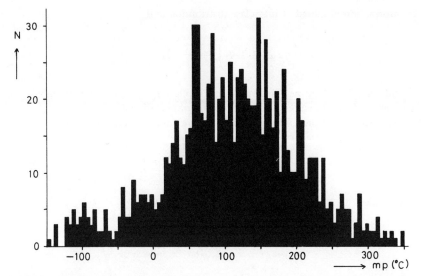

Figure 4.6. Histogram for melting distribution of 1001 arbitrarily chosen compounds. Class width 5°C; range −150 to +350°C.

provides more information. In order to account for this effect we assume an accuracy for the melting points equal to k times that of the histogram intervals. The m.p. measurement may be considered as a combination of two experiments:

1. A class selection procedure with an information yield of $I_1 = -\sum_{i=1}^{n} p_i \,\mathrm{ld}\, p_i$.
2. Within a class selection of one possibility out of a set of k, with an information yield of $I_2 = \mathrm{ld}\, k$, or in terms of class broadness Δx and reliability interval Δy $I_2 = \mathrm{ld}\, \Delta x / \Delta y$.

The total amount of information thus becomes

$$I = -\sum_{i=1}^{n} P_i \,\mathrm{ld}\, P_i + \mathrm{ld}\,\frac{\Delta x}{\Delta y} \qquad (4.31)$$

This calculation implies equal probability for all k possibilities within a given class. In many situations this is not true. In order to deal with situations where the probability of the situations are not equal, a probability distribution function within a class may be defined.

The equations involved in such situations are

$$I_1' = -\sum_{i=1}^{n} P_i \,\mathrm{ld}\, P_i \qquad (4.32)$$

for information obtained of intraclass distribution and

$$I_2' = \left(-\sum_{j=1}^{k} q_j \operatorname{ld} q_j \right)_i = \left(-\sum_{j=1}^{\Delta x/\Delta y} q_j \operatorname{ld} q_j \right)_i \tag{4.33}$$

For information of intraclass distribution in class i, q_j equals the probability of occurrence of the measured melting point in graph j of class i.

The average intraclass information $\bar{I}_2'$ for the entire histogram equals

$$\bar{I}_2' = \frac{1}{n} \sum_{i=1}^{n} \left(-\sum_{j=1}^{\Delta x/\Delta y} q_j \operatorname{ld} q_j \right)_i \tag{4.34}$$

which makes the total information $I_{\mathrm{tot}} = I_1' + \bar{I}_2'$

$$I_{\mathrm{tot}} = -\sum_{i=1}^{n} P_i \operatorname{ld} P_i + \frac{1}{n} \sum_{i=1}^{n} \left(-\sum_{j=1}^{\Delta x/\Delta y} q_j \operatorname{ld} q_j \right)_i \tag{4.35}$$

In the special situation where q_i has a constant value $\Delta y/\Delta x$, this equation simplifies to

$$I_{\mathrm{tot}} = -\sum_{i=1}^{n} P_i \operatorname{ld} P_i + \operatorname{ld} \frac{\Delta x}{\Delta y}$$

$$= -\sum_{j=1}^{m} P_j' \operatorname{ld} P_j' \tag{4.36}$$

with $m = n\Delta x/\Delta y$ and $P_j' = \Delta x$.

This equation refers to the distribution of melting point over a new histogram with class width m.

Based on the histogram and an accuracy of 1°C in the measurement of the melting points an amount of information may be obtained

$$I_{\mathrm{tot}} = 6.17 + \operatorname{ld} \frac{5}{1} = 8.5 \text{ (bit)}$$

Now one may ask how to use this information and whether this is enough? Based on an estimated number of 10^6 compounds, the identification of one particular compound requires an amount of information $I = \operatorname{ld} n_0/n_1$, which means here $\operatorname{ld} 10^6/1 = 20$ bits. The amount of information obtained from the melting point, namely, 8.5 bits, reduces the number of possibilities by $2^{8.5} \approx 360$, which means that a selection has to be made out of 300 remaining possibilities.

Another useful identification method is thin-layer chromatography (MA 73). Information theory can be applied to this method in order to find the best separation and thus the maximal possible identification. This may be illustrated by an example. In order to identify one particular compound out of a set of 8, an amount of information of $\operatorname{ld} 8 = 3$ bits is required.

TABLE 4.8. R_F **-VALUES OF EIGHT EXAMPLES FOR THREE DIFFERENT SEPARATING LIQUIDS (MA 73)**

Compound	Liquid 1	Liquid 2	Liquid 3
a	0.20	0.20	0.20
b	0.20	0.40	0.20
c	0.40	0.20	0.20
d	0.40	0.40	0.20
e	0.60	0.20	0.40
f	0.60	0.40	0.40
g	0.80	0.20	0.40
h	0.80	0.40	0.40
Amount of information	2 bits	1 bit	1 bit
$-P_i \mathrm{ld}\, P_i$	$(\log_2 4)$	$(\log_2 2)$	$(\log_2 2)$
	$-4 * \frac{1}{2} \mathrm{ld}\, \frac{1}{2}$	$-2 * \frac{1}{4} \mathrm{ld}\, \frac{1}{4}$	

Reproduced from *J. Chromatogr.*, **79**, 159 (1973) with permission of Elsevier Scientific Publishing Company, Amsterdam.

A theoretical solution for this identification problem is provided by two separating schemes, one giving 2 bits information and another 1 bit. In practice this solution is not always possible because part of the information provided by the first solvent may be identical with that given by the second solvent, which results in a total gain of information less than 3 bits. Therefore most information is obtained from a combination of two separating solvents with the highest possible independent information yield.

This can be illustrated by the example given in Table 4.8. It is concluded from the R_F-values that the individual liquids do not provide enough information for the identification of the compounds *a–h*. Thus the amount of information is enhanced by application of two successive separations with different separating liquids. There are various combinations of liquids possible. From Figure 4.7 one concludes that the combination of L_1 with L_3 provides 2 bits information only,

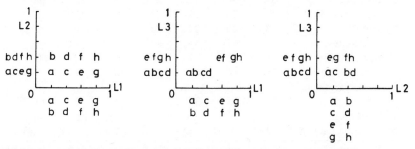

Figure 4.7. Schematic representation of two successive separations with different separating liquids L_1, L_2, and L_3 for three different selections of liquids.

TABLE 4.9. Information Loss by Correlation

Combination of Separating Liquids	Correlation[a] Coefficient	I_1	I_2 (bits)	I_3	I_{tot}
L_1/L_2	0	2	1		3
L_1/L_3	0.9	2		1	2
L_2/L_3	0		1	1	2

[a]$\rho_{i,j} = \sigma_{i,j}/\sigma_i, \sigma_j$.

because the information provided by L_3 has been given already by L_1. The same conclusion holds for the combination L_2 and L_3. The only combination that provides all information required for identification is L_1 with L_2.

Selecting appropriate separating liquids for separation of larger members of compounds follows the same line in principle. Mostly the interdependence of the information provided follows from the ratio of R_F values of compounds for the different separating liquids. When the R_F values obtained from two different separating liquids are highly correlated, the information gain in combined separation is small and no extra separation is obtained. A low correlation provides better separation and more information on application of two separations. This is demonstrated in Table 4.9.

Apparently I_{tot} is less than $I_i + I_j$ on high correlation between i and j.

TABLE 4.10. $100 * R_F$-Values and Information Gain of Cartenoids with Seven Separating Liquids of Varying Polarity (MA 73)

Compound	Separating Liquids Ordered in Increasing Polarity						
	1	2	3	4	5	6	7
Cryptoxanthine	62	70	76	74	39	21	4
Rubixanthine	45	60	64	45	15	4	0
Lycoxanthine	29	37	40	32	8	0	0
Isozeanthine	34	56	92	91	57	36	10
E scholtzanthine	12	22	25	22	8	1	0
Lycophyll	8	20	22	20	7	0	0
Euglenanone	62	68	81	80	54	34	9
Canthaxanthine	58	65	79	80	55	37	20
Rhodoxanthine	28	42	43	40	14	7	1
8′-Apo-β-carotenic acid	28	38	30	15	5	0	0
Torularhodin	6	10	9	2	1	0	0
Information gain (bit)	2.66	2.91	2.91	3.09	2.11	1.79	1.49

Reproduced from *J. Chromatogr.* **79** 159 (1973) with permission of Elsevier Scientific Publishing Company, Amsterdam.

TABLE 4.11.   Linear Correlation Coefficient ρ^a Between R_F-Values of Various Separating Liquids

	1	2	3	4	5	6	7
1	1.00	.97	.86	.82	.77	.74	.63
2		1.00	.94	.89	.81	.76	.62
3			1.00	.98	.93	.89	.72
4				1.00	.97	.94	.78
5					1.00	.99	.87
6						1.00	.90
7							1.00

$$^a\rho_{i,j} = \frac{\sigma_{i,j}}{\sigma_i * \sigma_j}$$

$$\sigma_x = \left\{ \sum_{x=-\infty}^{x=+\infty} [x - \varepsilon(x)]^2 p(x) \right\}^{1/2}$$

$$\sigma_{i,j} = \varepsilon\{[i - \varepsilon(i)][j - \varepsilon(j)]\}$$

Table 4.10 gives R_F-values and information yields of a number of compounds and liquids used in thin-layer chromatography. The polarity of separating liquids influences the distribution of the R_F-values. Liquids with like polarities have highly correlated R_F-values. In calculating the information gain it has been assumed that R_F-values differing up to three units in Table 4.10 (0.03 R_F unit) do not produce a separation. Compounds with R_F-values that differ up to 0.03 are found in the same class.

In order to calculate the information gain on application of two separating liquids consecutively, the correlation between the separating capability of these two liquids must be known. In Table 4.11 the linear correlation coefficients $\rho_{i,j}$ between R_F-values of separating liquids i and j are listed.

The information gain of a combination of two separating liquids is found by application of the Shannon equation

$$I_{1,2} = - \sum_i \sum_j P(i,j) \operatorname{ld} P(i,j) \tag{4.37}$$

Here $P(i, j)$ denotes the probability of finding an R_F-value with liquid 1 in class i and with liquid 2 in class j. Here too it has been assumed that R_F values that differ up to 0.03 do not produce a separation (see Table 4.12).

- The information required for identification of one compound out of 11 equals $\operatorname{ld} 11 = 3.46$ (bits).
- The total amount of information obtained depends on the sum of information of the individual separating liquids, corrected for their

TABLE 4.12. Information Gain for All Combinations of Two Separating Liquids

	1	2	3	4	5	6	7
1	2.66	3.10	3.28	3.46	3.46	3.10	2.85
2		2.91	3.28	3.28	3.28	3.10	3.10
3			2.91	3.10	3.10	3.10	3.10
4				3.10	3.10	3.10	3.28
5					2.11	2.41	2.66
6						1.79	2.04
7							1.49

mutual correlation:

$$I_{tot} = F(I_1, I_2, \rho_{1,2}) = 3.46 \text{ bit (Table 4.12)}$$

Using the combinations 1–4 or 1–5 it is possible to separate all components. The mathematical expression for I_{tot} is used to calculate the information gained by gas chromatography. Here it is assumed that the distribution of retention indexes may be described by a Gaussian curve, which provides the information

$$I = - \int P_m(x_i) \operatorname{ld} P_m(x_i)\, dx_i + \int P_e(x_i) \operatorname{ld} P_e(x_i)\, dx_i \qquad (4.38)$$

Here $P_m(x_i)$ denotes the distribution of retention indexes as given in the literature and $P_e(x_i)$ gives the accuracy of the measured retention index (Figure 4.8).

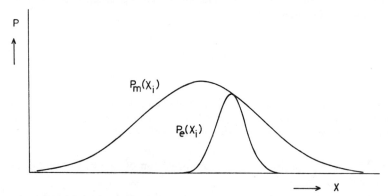

Figure 4.8. Distribution of retention index $P_m(x_i)$ and accuracy of measured retention index $P_e(x_i)$.

The information content of a column type i equals

$$I_i = \tfrac{1}{2} \mathrm{ld} \frac{\sigma_m^2}{\sigma_e^2}$$

in which σ_m^2 equals the variance of function P_m and σ_e^2 that of function P_e. It is seen from the literature that the information content of most column types that are used in practice is about the same and between 6.5 and 7 bits. A combination of two or more columns may give quite different amounts of information, depending on the combination one selects.

$$I_{1,2,3,\dots,n} = -\int_1\int_2 \cdots \int_n P_m(x_1 \cdots x_n)\,\mathrm{ld}\,P_m(x_1 \cdots x_n)\,dx_1\,dx_2 \cdots dx_n$$

$$+ \int_1\int_2 \cdots \int_n P_e(x_1 \cdots x_n)\,\mathrm{ld}\,P_e(x_1 \cdots x_n)\,dx_1\,dx_2 \cdots dx_n$$

$$= \tfrac{1}{2}\,\mathrm{ld}\,\frac{|\mathrm{cov}_m|}{|\mathrm{cov}_e|} \tag{4.39}$$

Here $P_m(x_1 \cdots x_n)$ denotes an n-dimensional Gaussian distribution and $|\mathrm{cov}_m|$ equals the determinant of the covariance matrix:

$$|\mathrm{cov}_m| = \begin{vmatrix} \sigma_1^2 & \sigma_{1,2} & \cdots & \sigma_{1,n} \\ \sigma_{2,1} & \sigma_2^2 & \cdots & \sigma_{2,n} \\ \vdots & & & \\ \sigma_{n,1} & \cdots & \cdots & \sigma_n^2 \end{vmatrix} \tag{4.40}$$

By substitution of the correlation

$$\rho_{i,j} = \frac{\sigma_{i,j}}{\sigma_i * \sigma_j} \tag{4.41}$$

one gets

$$|\mathrm{cov}_m| = \sigma_{m,1}^2 * \sigma_{m,2}^2 * \sigma_{m,3}^2 \cdots \sigma_{m,n}^2 * \begin{vmatrix} 1 & \rho_{1,2} & \rho_{1,3} & \cdots & \rho_{1,n} \\ \rho_{2,1} & 1 & . & \cdots & \rho_{2,n} \\ \vdots & & & & \\ \rho_{n,1} & & & & 1 \end{vmatrix} \tag{4.42}$$

In the expression for $|\text{cov}_e|$ all terms $\rho_{i,j}$ in $|\text{corr}|$ are equal to zero because of the noncorrelation of measuring errors in different columns; therefore

$$|\text{cov}_e| = \sigma_{e,1}^2 * \sigma_{e,2}^2 * \sigma_{e,2}^2 * \cdots \sigma_{e,n}^2 \tag{4.43}$$

Substitution of the relation for $|\text{cov}_m|$ and $|\text{cov}_e|$ in eq. 4.39 yields

$$
I_{1,2,\ldots,n} = \tfrac{1}{2}\,\text{ld}\left[\frac{\sigma_{m,1}^2 * \sigma_{m,2}^2 * \cdots \sigma_{m,n}^2}{\sigma_{e,1}^2 * \sigma_{e,2}^2 * \cdots \sigma_{e,n}^2} * \begin{vmatrix} 1 & \rho_{1,2} & \cdots & \rho_{1,n} \\ \rho_{2,1} & 1 & & \\ \vdots & & & \\ \rho_{n,1} & & & 1 \end{vmatrix}\right]
$$

$$
= \sum_{i=1}^{n}\left(\tfrac{1}{2}\,\text{ld}\,\frac{\sigma_{m,i}^2}{\sigma_{e,i}^2}\right) + \tfrac{1}{2}\,\text{ld}|\text{corr}|
$$

$$
= \sum_{i=1}^{n} I_i + \tfrac{1}{2}\,\text{ld}|\text{corr}| \tag{4.44}
$$

The values of $\rho_{i,j}$ and $\sigma_{m,i}$ are known from the literature; the value of $\sigma_{e,i}$ equals 2, a value that is known from practice. This enables an estimation of I for each combination of n columns. In order to identify a compound with the minimum number of columns one should select the combination that offers the maximum amount of information. Based on this consideration the order of columns found in the McReynolds table reads as shown in Table 4.13.

TABLE 4.13. Optimal Order of Columns (DP 75)

Number	Stationary Phase	Information per Column (bit)	Total Information (bit)
1	Diglycerol	7.0	7.0
2	Squalane	6.6	13.2
3	Diethylglycol succinate	7.0	18.8
4	Carbowax 20 m	6.8	24.1
5	Poly(phenyl ether-6-ring)	6.7	28.6
6	Tricresyl phosphate	6.7	32.3
7	Apiezon L	6.6	35.3
8	Diisodecyl phthalate	6.6	38.1
9	S.E. 30	6.6	40.7
10	Bis(ethoxyethyl) phthalate	6.7	43.3

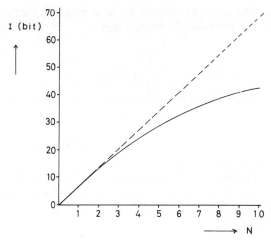

Figure 4.9. Information gain I as a function of the number of columns N with (——) and without (-----) correlation. Columns are ordered according to Table 4.13.

In Figure 4.9 the information content is given as a function of the number of columns.

4.3 DATA REDUCTION

In order to apply measured quantities, in most situations the following treatment is required:

- Removal of nonessential—irrelevant data—(noise).
- Reconstruction of essential data.
- Interpretation of data.

These treatments are discussed in the next sections.

4.3.1 Removal of Irrelevant Data (Noise)

A practical definition of noise is the undesired part of a signal. In practice it is impossible to separate noise completely from the rest of the signal because each noise filter generates noise itself. A practical yardstick for the applicability of a signal is provided by the signal-to-noise ratio S/N. In order to measure this ratio one starts with measuring the available power of the signal wave and of the noise wave. In direct current signals the power is proportional to the voltage E; in alternating current signals it is proportional to the root mean square value $A_{\text{r.m.s.}}$ of the wave shape:

$$A_{\text{r.m.s.}} = \left[\sum_{i}^{K} \frac{(\bar{A} - A_i)^2}{K} \right]^{1/2} \tag{4.45}$$

with $\bar{A}$ the average value of the wave shape and A_i the deviation of the ith measurement from the average.

With direct current signals sampled at discrete time intervals the main part of the noise is governed by a variation with time of the signal. Here the noise is given by the standard deviation of the signal and thus the signal-to-noise ratio becomes

$$\frac{S}{N} = \frac{\text{average value}}{\text{standard deviation}} = \frac{1}{\text{relative standard deviation}} \tag{4.46}$$

Equation 4.46 gives a simple method to calculate S/N from a recorder signal; for example, 99% of all observations is found between $\mu - 2.5\sigma$ and $\mu + 2.5\sigma$ (μ equals the main value and σ denotes the estimated standard deviation). Here S/N equals 0.2 (HE 72), the peak-to-peak variance of the signal. The observation period should be long enough to enable a reliable estimate of S. It is assumed that the signal is distributed around the average value μ according to a Gaussian distribution.

In situations where alternating current signal are measured, other techniques should be applied for calculating S/N, for example, autocorrelation (see Section 4.5.2).

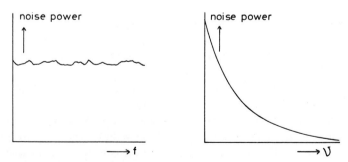

Figure 4.10. Frequency spectrum of "white" noise and of "$1/f$" noise.

One may distinguish various types of noise that can be recognized by their frequency spectrum Figure 4.10:

- White noise has a frequency spectrum with a flat shape. It is caused by random processes such as Brownian motion, external influence, and thermal motion of electrons (Johnson noise).
- $1/f$ noised. Here the power is proportional to $1/f$. Such a situation is met with, for example, a drifting signal.
- Red noise (contains more low-frequency noise than the average value).
- Blue noise (contains more high-frequency noise than the average value).
- Interference noise. This noise nearly always results from external sources with relatively low chance of occurrence, such as switching on or off of power, undesired components in a spectrum, and so on. A special form of interference noise is impulse noise. It is caused by sources such as starting up a powerful engine. All frequencies are met in this type of noise, which makes it difficult to eliminate it. Another type of interference noise is deformation noise. Here some specific frequencies are added to or removed from the signal. Some possible causes for deformation noise are nonlinearity of the instrument, transmission losses, and reflections.

4.3.2 Data Filtering

In practice all types of noise occur and give a mixture. The noise energy generally depends on $1/f^n$ where $0 < n < 1$.

Removal of noise from a signal is performed by filtering. In order to construct a good filter, one starts with fixing the desired frequency or frequencies; based on these, a filter is designed that allows passage of frequencies in a very narrow range around the selected one.

Analogue filtering operates on voltages, provided by a measuring instrument. In some types of modern instruments the measured quantities (voltages) are converted internally by an analogue to digital (AD) convector and presented as a digital number or a combination of bits on a magnetic device. Here digital filtering should be applied for the removal of noise. Apart from this, some physical methods, such as radioactivity measurement, provide the results in a digital form without the need for AD conversion.

A dc signal is filtered by allowing only low frequencies to pass. This is done by means of smoothing: the signal is integrated over an extended period in order to cancel the higher frequencies. Analogue smoothing is

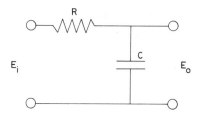

Figure 4.11. Simple analogue smoothing filter with resistance R and capacitance C producing an output signal E_o from an input signal E_i.

performed by RC filters. The input voltage E_i is supplied to a resistor R parallel to a capacitor C and thus produces the output voltage E_o (Figure 4.11). The output signal as a function of time $(E_o)_t$ equals

$$(E_o)_t = (RC)^{-1} \int_0^t (E_i)_t \, dt \tag{4.47}$$

Digital Filtering

Digital smoothing asks for the running average of a series of measurements x_t as a function of time.

$$S_t = \frac{1}{N}(x_t + x_{t-1} + \cdots + x_{t-N+1}) \tag{4.48}$$

or

$$S_t = S_{t-1} + \frac{1}{N}(x_t - x_{t-N}) \tag{4.49}$$

in which S_t equals the running average at time t and N the time span of the observation period expressed in time units.

Equation 4.49 describes how to calculate a new S_t value from an older S_{t-1} by adding a new measured quantity x_t and eliminating the oldest x_{t-N}. This procedure asks for greater memory capacity because a complete series of N data should be stored.

Exponential smoothing requires considerably less storage capacity because nonavailable data are substituted by the mean value: in eq. 4.49 $(x_t - x_{t-N})/N$ is substituted by $(x_t - S_{t-1})/N'$. Here S_t equals the running exponential mean at time t and N' the mean length of the averaging period. Substitution of $\alpha = N'^{-1}$ in eq. 4.49 yields

$$S_t = \alpha x_t + (1-\alpha)S_{t-1}$$

$$= S_{t-1} + \alpha(x_t - S_{t-1}) \tag{4.50}$$

It can be seen that the new S_t value is obtained from the old by correcting with the scaled difference between the old one and the most recent measured value. The scaling factor α is inversely proportional to the length of the running interval. The advantage of this method over the previous one is that only the mean value S_{t-1} has to be stored in memory; moreover, all measured values are accounted for, in contrast to the running average method. The importance of a measurement decreases with its age; this is of consequence for the mean age $\bar{k}$ of a series of measurements (BR 62). The age of the most recent one equals zero, the next to the most recent one 1, and so on. Thus the average age $\bar{k}$ equals

$$\bar{k} = \frac{1-\alpha}{\alpha} \tag{4.51}$$

The running average method has an average age of the observations for a period N

$$\bar{k} = \frac{0+1+2+3+\cdots+(N-1)}{N} = \frac{N-1}{2} \tag{4.52}$$

Equalizing these two ages yields

$$\frac{1-\alpha}{\alpha} = \frac{N-1}{2} \tag{4.53}$$

or

$$\alpha = \frac{2}{N+1} \tag{4.54}$$

The weight factor depicted as a function of time is given in Figure 4.12.

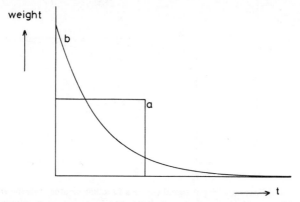

Figure 4.12. Weight factor as a function of time for running average (a) and for exponential smoothing (b).

The shape of a signal is altered by smoothing and depends on the smoothing technique applied. It is seen in Figure 4.13 that the shape of the signal depends on the smoothing procedure irrespective of whether an equal time span for the observations is applied.

Frequency filtering for one specific frequency ν_0 allows only ν_0 to pass the filter. All time-dependent changes of the signal are lost. Therefore the filter should allow a well-defined bandwidth. In order to prevent an undesired quantity of noise to pass through this band the signal is transferred to a frequency region that has a low amount of natural noise. Mostly this is a high-frequency region and the signal of interest is superimposed upon a high-frequency signal. This can be very useful in eliminating $1/f$ noise and/or 50 Hz noise (or 60 Hz). The superimposing procedure is called modulating.

In photometric applications chopping procedures are applied. Here an ac signal is added to the dc measuring signal followed by a filtering at the chopping frequency to eliminate $1/f$ noise.

Nonperiodic but repeatable signals can be noise filtered by averaging (HI 72). A simple example is given by gravimetric analysis. Repeating the analysis provides a decrease of the noise according to $n^{-1/2}$ ($n=$ number of repetitions). Signal series should be sampled with a frequency at least twice the highest occurring frequency in order to describe the signal series completely (see Fourier Convolution below). In order to average signal series in a well-defined way synchronization should be applied. The synchronization pulse should be deferred from the signal itself or should trigger the signal registration. This method has been elegantly applied by

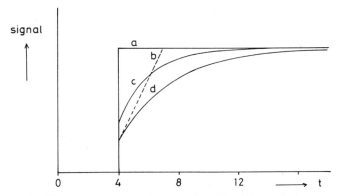

Figure 4.13. Modification of input signal (a) as a function of time with the running average method with $N=4$ (b), with an exponential with $\alpha=\frac{1}{4}$ (d), and with an exponential $\alpha=2/N+1=0.4$ (c).

Smit (SM 70, SM 80), who describes a method for pseudorandom injections of a sample into a chromatographic column and cross-correlating the detector signal with the noise generator that governs the injections. With this method it is possible to accumulate a signal from a detector output that under normal conditions is hidden in the background noise.

A method for producing synchronization pulses from the signal itself consists of monitoring the sign of the derivative and selecting the moments where the sign equals zero after passing from negative toward positive values.

Digital signal averaging is a commonly used technique that requires as many channels as signals to be averaged. In order to avoid this sometimes a boxcar integrator can be used, which requires one channel only. Here the integral boundary is stepwise altered for signal monitoring. Interference noise is not averaged out by signal averaging.

In situations where synchronization pulses are not available, correlation techniques can be useful. Here the signal value at a moment t is multiplied with a signal value an interval τ apart; thus

$$\varphi_{zz}(\tau) = \sum_{t=1}^{N-\tau} \frac{(z_t - \bar{z}_t)(z_{t+\tau} - \bar{z}_{t+\tau})}{(n-\tau)\sigma_z^2} \tag{4.55}$$

(for a stationary signal $\bar{z}_t = \bar{z}_{t+\tau}$).

The correlated signals become clearly distinguishable because noise is eliminated. The correlation time may be used to describe the signal. Autocorrelating a sine function produces a cosine function. It is known that $\tau = 0$ produces the variance σ^2 of the signal; it consists of signal and noise.

Kalman Filtering

A measured quantity w_t, obtained at time t, is just a momentary description of a situation and pays no attention to the history of a time series of measurements. However, it is possible to make a better estimate of reconstructed values out of a time series at time t and to predict the value of new measurements taking into account the history of that time series. Here the prediction of a measured value is based on the most recent measurement and the expected value is corrected with the values measured in the more distant past. The weight of the predicting capacity of the measured values from the past decreases with increasing distance from the present time. The application of this technique is called exponential smoothing. Kalman (KA 61) showed that there exists an optimal estimation method that is even better than the exponential smoothing method. Analytical chemists have recently become aware of this technique (SE 76). The application is limited to first-order autoregressive stochastic stationary processes with a time constant T_x. For such processes the predicted value for time t, called $\bar{x}_t$, is combined with a measured value w_t in order to construct a better estimation named $\hat{x}_t$ (each of the data x_t and w_t contains errors, which makes an exact prediction impossible):

$$\hat{x}_t = (1 - k_t)\bar{x}_t + k_t w_t \qquad (4.56)$$

in which $\bar{x}_t = \exp(-1/T_x)\hat{x}_{t-1}$.

The reconstruction $\hat{x}_t$ that can be found from this linear combination depends on the value of k_t. The formulation of the linear equation is such that the reconstruction gives a pure estimation $w_k = x_k + v_k$, v_k being measurement noise. This means that the average deviation of the reconstructed value from the true value equals zero. The value of k_t defines the weight of the predicted value with respect to the weight of the measured value. The optimal value of k_t results from a consideration of the variances of $\hat{x}_t$ and w_t which are assumed to be independent. The variance of $\hat{x}_t$, called $\mathrm{var}(\hat{x}_t)$, can be written as

$$\mathrm{var}(\hat{x}_t) = (1 - k_t)^2 \mathrm{var}(\bar{x}_t) + k_t^2 \mathrm{var}(w_t) + 2k_t(1 - k_t)\mathrm{cov}(\bar{x}_t, w_t)$$
$$= (1 - k_t)^2 \mathrm{var}(\bar{x}_t) + k_t^2 \mathrm{var}(w_t) \qquad (4.57)$$

the covariance between $\bar{x}_t$ and w_t being zero because of their mutual independence. Now the best k_t value is found from a minimalization of $\mathrm{var}(\hat{x}_t)$:

$$\frac{\partial \mathrm{var}(\hat{x}_t)}{\partial k_t} = 0 = -2(1 - k_t)\mathrm{var}(\bar{x}_t) + 2k_t \mathrm{var}(w_t) \qquad (4.58)$$

or the optimal value for k_t equals

$$k_t = \frac{\mathrm{var}(\bar{x}_t)}{\mathrm{var}(\bar{x}_t) + \mathrm{var}(w_t)} \qquad 0 \leqslant k_t \leqslant 1 \qquad (4.59)$$

Substitution of this k_t-value in eq. 4.57 yields

$$\mathrm{var}(\hat{x}_t) = k_t \mathrm{var}(w_t)$$
$$= (1 - k_t)\mathrm{var}(\bar{x}_t) \qquad (4.60)$$

Because $0 \leqslant k_t \leqslant 1$, the variance in the reconstructed value $\mathrm{var}(\bar{x}_t)$ lies between $\mathrm{var}(\bar{x}_t)$ and $\mathrm{var}(w_t)$, which is most desirable. The variance in $\bar{x}_t$ equals

$$\mathrm{var}(\bar{x}_t) = \exp(-2/T_x)\mathrm{var}(\hat{x}_{t-1}) + [1 - \exp(-2/T_x)]\mathrm{var}(x_t) \quad (4.61)$$

in which $\mathrm{var}(x_t)$ represents the variance of the entire process under consideration—a stochastic, stationary, autoregressive process with a time constant T_x.

The recursive relations that enable an estimate of process values and the reliability of the predictions from a limited process realization period can be summarized as

$$\bar{x}_t = \exp(-1/T_x)\hat{x}_{t-1} \qquad\qquad\qquad (4.62)$$

$$\hat{x}_{t-1} = (1 - k_{t-1})\bar{x}_{t-1} + k_{t-1}w_{t-1} \qquad\qquad (4.63)$$

$$\mathrm{var}(\bar{x}_t) = \exp(-2/T_x)\mathrm{var}(\hat{x}_{t-1}) + [1 - \exp(-2/T_x)]\mathrm{var}(x_t) \quad (4.64)$$

$$\mathrm{var}(\hat{x}_t) = k_t\mathrm{var}(w_t) \qquad\qquad\qquad (4.65)$$

and

$$k_t = \frac{\mathrm{var}(\bar{x}_t)}{\mathrm{var}(\bar{x}_t) + \mathrm{var}(w_t)} \qquad\qquad\qquad (4.66)$$

The difference between exponential smoothing and Kalman filtering can be summarized as follows: exponential smoothing reconstructs from a combination of measurements and predictions, whereas Kalman also includes the autocorrelation function in its reconstruction.

Prediction over a period v yields

$$\bar{x}_{t+v} = \exp(-v/T_x)\hat{x}_t \qquad\qquad\qquad (4.67)$$

$$= \exp[-(v+1)/T_x](1 - k_t)\hat{x}_{t-1} + k_t w_t \exp(-v/T_x)$$

and

$$\mathrm{var}(\bar{x}_{t+v}) = \exp(-2v/T_x)\mathrm{var}(\hat{x}_t) + \mathrm{var}(x_t)[1 - \exp(-2v/T_x)]$$

$$(4.68)$$

Müskens (MS 78) applies a one-dimensional Kalman filtering method with constant process and analysis parameter to develop and describe an analysis scheme for monitoring the ammonia concentration of the Rhine River.

Here the prediction of the ammonia concentration on days in the future is based on the known characterization of the autocorrelation function of the ammonia content of the river water by an exponential function with a given time constant. The variance var(x_t) is also known from historical data. The starting parameter values in general follow from the model of the process to be reconstructed. The average ammonia concentration over a year and the variance can be used as starting values for the Kalman algorithm.

In order to test the capabilities of the procedure for the prediction of concentrations in the future, the standard deviation of the difference between measured and predicted values is compared with that of the difference between measured values at a time $t+v$ in the future and those measured at time t.

The latter situation describes a situation where a measured concentration at time t is considered to be constant till $t+v$, the best one can do without any predicting strategy such as application of a hold circuit. In Figure 4.14 these standard deviations are plotted as a function of the prediction period, denoted as sampling

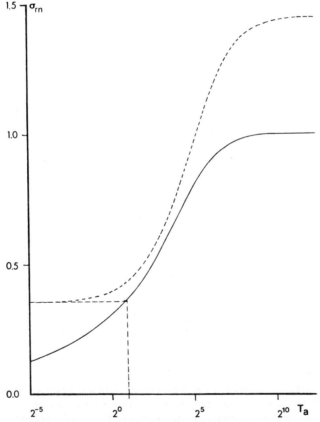

Figure 4.14. Reconstruction error σ_{rn} as a function of sampling interval, T_a; using filtering (——) and without filtering (-----), using a hold circuit.

interval T_a in days. It is seen that the hold circuit method (no filtering applied) produces the same standard deviation as the Kalman prediction over a period of about 5 days. The implication of this observation in practice is that 80% of the analyzing effort can be saved without a significant loss of information. Instead of one NH_4^+ analysis per day, one each fifth day suffices.

Application of the Kalman procedure to a four-dimensional filtering problem is given by Poulisse (PO 79). He uses a multicomponent analysis computation on spectroscopic data and compares this method of calculation with a nonrecursive least-squares estimation method (see Section 4.4). His calculations result in a determination of the number of components in the mixture, an estimation of a constant systematic error, and a computation of the error covariance matrix. In this article theory and practice indicate a new design of measurement experiments and novel applications of on-line computation and computer control. It is shown that the Kalman filter algorithm, because of its recursive nature and computational efficiency, is a very convenient method for multicomponent analysis computations and offers promising possibilities in acquiring a predetermined quality of the results in analytical chemistry.

Another application of Kalman filtering is presented by Didden and Poulisse (D1 80). Here a Kalman filter is designed to determine the number of components in multicomponent mixtures. Optical spectra (UV–Vis) of different mixtures are simulated and processed sequentially by the Kalman filter. It is shown that the number of components present in the sample as well as their concentrations can be determined simultaneously. The advantages of the method in view of the computational effort required are demonstrated.

Fourier Convolution

Here a time series is written as a summation of a number of sine and cosine terms. This procedure is allowed for each type of function $f(x)$:

$$f(x) = a_1 \sin x + a_2 \sin 2x + a_3 \sin 3x + \cdots + b_0 + b_1 \cos x + \cdots + b_n \cos nx$$

$$(4.69)$$

In order to calculate the coefficients a_m, $f(x)$ is multiplied by $\sin mx\, dx$ $(m = 1, 2 \ldots)$ and integrated between $-\pi$ and π. This process is visualized as shown in Figure 4.15. It is seen that even functions "$f(x) = +f(-x)$" make all a coefficients zero and odd functions "$f(x) = -f(-x)$" make all b coefficients zero.

In situations where the periodicity of $\sin mx$ corresponds with a periodicity of $f(x)$, the product function does not change sign because a change in sign of $\sin mx$ occurs at the same x-value where a change in sign of $f(x)$ occurs. All periods of the product function contribute with the same algebraic sign to the integral and the result is a nonzero value of a_m. In situations where the periodicity of $\sin mx$ does not correspond with a

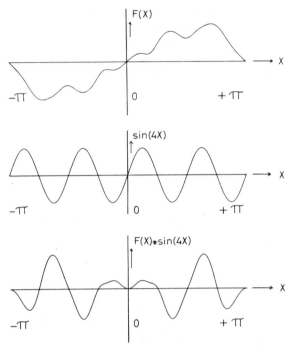

Figure 4.15. Schematic representation of the convolution of an odd function $F(x)$ with the sine function $\sin(4x)$.

periodicity of $f(x)$ the product function tends to be $+$ as often as $-$ and the contributions of the periods to the integral tend to cancel out each other. The result is a practical zero value for a_m.

The same procedure is repeated with $\cos mx$ $(m=0,1,2,\dots)$ in order to find coefficients b_m.

Generally a series contains a limited number of terms. Where the highest frequency of the series equals mx, the base frequency consists of m terms, and thus the series consists of m cosine functions, m sine functions, and one constant, which means $2m+1$ terms. The amplitudes are derived from $2m+1$ measurements, which means that the number of measurements (minimal and in practice also maximal) equals about $2m$.

The sampling has to provide $2m$ samples which at a constant sampling frequency v means an intersample time span Δt:

$$\Delta t = \frac{v}{2m} = \frac{v}{2v} \tag{4.70}$$

v equals the length of the series expressed in time units. The shape of a

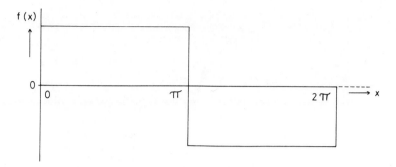

Figure 4.16. Representation of an odd square wave function.

wave may be characterized by a number of sine and cosine functions (HO 71). An even function requires a description with exclusively cosine functions, for example, $f(x) = -\frac{1}{4}\pi$ with $0 < x < \pi$; $f(x) = 0$ with $x = 0$ and $f(x) = \frac{1}{4}\pi$ with $-\pi < x < 0$. Here $f(x)$ is an odd function and the series is formed by a summation of sine terms exclusively (Figure 4.16).

$$f(x) = a_1 \sin x + a_2 \sin 2x + a_3 \sin 3x \cdots \qquad (4.71)$$

$$a_m = \frac{2}{\pi} \int_0^\pi f(x) \sin mx \, dx \qquad m = 0, 1, 2, 3, \ldots$$

$$= \frac{1}{2} \int_0^\pi \sin mx \, dx$$

$$= -\frac{1}{2m} \cos mx \bigg]_0^\pi \qquad (4.72)$$

m odd$\rightarrow a_m = 1/m$
m even$\rightarrow a_m = 0$
Thus $0 < x < \pi$ yields

$$\tfrac{1}{4}\pi = \sin x + \tfrac{1}{3}\sin 3x + \tfrac{1}{5}\sin 5x \cdots \qquad (4.73)$$

See Figure 4.17. The frequency spectrum of $f(x)$ is shown in Figure 4.18.

The general description of a function by means of a summation of sine and cosine function is

$$f(x) = \int_{-\infty}^{+\infty} f(\nu) e^{2\pi i x \nu} \, d\nu \qquad (4.74)$$

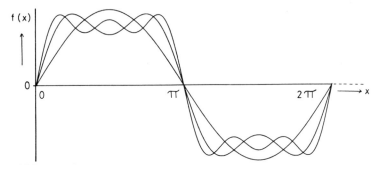

Figure 4.17. Construction of an odd square wave function by summation of sine functions (eq. 4.73).

and of its Fourier transform

$$f(v) = \int_{-\infty}^{+\infty} f(x)e^{-2\pi i x v}x \qquad (i = \sqrt{-1}) \tag{4.75}$$

$f(v)$ depends on the frequency and $f(x)$ on position or time x, or in other words $f(v)$ is defined in a frequency domain and $f(x)$ is the position or time domain. The conversion of the general equation to the sum of sine and cosine function is made by

$$e^{-2\pi i x v} = \cos(2\pi x v) - i\sin(2\pi x v) \tag{4.76}$$

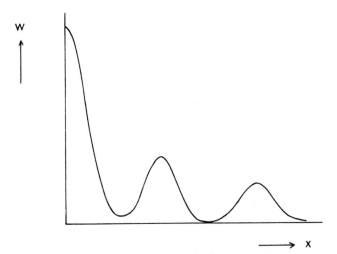

Figure 4.18. Frequency spectrum of $f(x)$ (eq. 4.73).

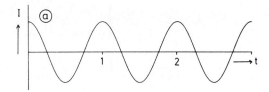

Figure 4.19. (*a*) A finite number of cosine periods (*b*) transformed with a Fourier transform.

The conversion of $f(x)$ into $f(\nu)$ is a rather time-consuming process which can be simplified considerably by application of a fast Fourier transform algorithm and a digital computer, provided the number of sample points in the series equals 2^m.

A hardware Fourier transform is provided by a spectroscopic prism. Here all frequencies of a given shape of wave form are converted into a spectrum amplitude as a function of frequency. This can be illustrated by considering a wave $f(x)$ formed by a finite series x of cosine periods of frequency ν', as shown in Figure 4.19a, where $y = \cos 2\pi x \nu'$. Transformation of y yields the Fourier transform $f(\nu)$ (Figure 4.19b).

$$f(\nu) = \frac{1}{2\pi(\nu' - \nu)} \sin\left[2\pi x(\nu' - \nu)\right] \qquad (4.77)$$

The width of $f(\nu)$ is inversely proportional to x (= the length of the finite time series).

In order to improve the S/N ratio of a signal, Fourier analysis can be applied, followed by convolution with an appropriate smoothing function, for example, an exponential function or a cutoff filter (Figure 4.20).

A smoothing filter acts as a filter that blocks high frequencies and allows only low-frequency signals to pass. Some examples of smoothing filters are a triangular function ($x = 0$, $y = 1$ and $x \geqslant a$, $y = 0$); the half of a Gaussian function ($x = 0$, $y = y_{max}$ and $x = a$, $y \rightarrow 0$); and a decreasing exponential function. The selection of an appropriate smoothing function is a matter of trial and error. Sometimes the characteristics of the instrument under

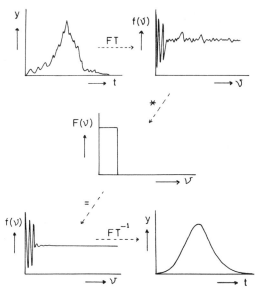

Figure 4.20. Improvement of S/N ratio by application of a Fourier transformation and a cutoff filter.

consideration are measured under high signal-to-noise conditions. These characteristics are used afterward in a convolution of signals measured under conditions with high noise levels. Here the spectrum including noise is convoluted by means of a Fourier transform with a virtually noise-free signal.

In the frequency domain convolution means multiplication; the place or time domain asks for a much more elaborate process for convolution. Here the convolution integral means

$$T(y) = \int_{-\infty}^{+\infty} t(x) B(y-x) \, dx \qquad (4.78)$$

in which $T(y)$ represents a virtually noise-free convoluted signal, $t(x)$ the observed signal, and $B(y-x)$ the smoothing function. A disadvantage of convolution should be mentioned here; that is, together with the noise part of the signal a distortion of the signal occurs which manifests itself as a broadening of the peaks (and a lowering of the maximal signal values because of normalization).

Deconvolution

Deconvolution offers a method of eliminating distortions of signals that are caused by the instrument. Here one aims for the recovery of the original signal, for example, by application of Fourier transform analysis. As an example the distortion of an optical signal caused by the slit of the instrument is considered. The shape of the distorted signal is known to be a broadened line and in order to recover the original line shape a transformation is required that performs the opposite of a convolution calculation:

$$T(y) = \int_{-\infty}^{+\infty} t(x)B(y-x)\,dx \qquad (4.79)$$

$T(y)$ represents the broadened signal, distorted by the Fourier transform (FT) of the slit width function $B(y-x)$. $t(x)$ equals the undistorted signal one wants to recover. The aim of the deconvolution process is the recovery of $t(x)$. In the frequency domain deconvolution means division of the FT of the observed signal by the FT of the distorting function and back-transformation of the resulting Fourier transform of the undistorted signal (HA 74).

The deconvolution process is depicted schematically in Figure 4.21. The deconvolution results in a smaller line width and thus allows a better peak separation. However, the deconvoluted signal shows some little extra signals that are not found in the original spectrum. These signals are artificial and hamper the detection of low-intensity signals that actually might be present in the original signal.

In situations where the slit function (or other artificial disturbances due to the instrument) is unknown it should be measured separately. A general

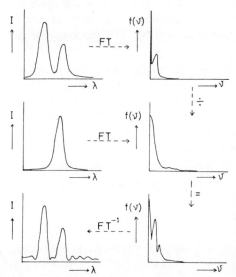

Figure 4.21. Removal of signal broadening by deconvolution of the Fourier transform of the distorted signal by the Fourier transform of the distorting signal followed by the reverse transformation.

procedure accepted in spectrometry is to proceed as follows: A spectrum of a component of a mixture is recorded with a narrow slit width A and another recording is made with identical parameter settings except for a large slit width B. It is to be noted that recording A gives a better peak separation than recording B. Now the Fourier-transformed spectrum B is divided by the FT of spectrum A in order to produce the FT of the slit function C'. The back-transformation of C' yields the slit function in the λ domain and may be used for transformation purposes in future spectra. It should be noted that this function is apparatus-bound and cannot be used for transformation of spectra produced by other instruments.

4.4   DATA RECONSTRUCTION AND CURVE FITTING

The production of measurements by one-dimensional methods of analysis such as melting points, boiling points, indexes of refraction, and so on, mostly do not present problems. A multiplication or division mostly suffices for reconstruction of relevant information. The situation becomes more complicated for two-dimensional methods of analysis such as encountered in spectral analysis. The analysis here means measurement of peak positions and of the heights or peak surface. Difficulties may be caused by the following:

- *Noise and Drift on the Original Spectra.* Noise may be eliminated by smoothing. Systematic deviations in the abscissa and ordinate (the λ and the intensity scale, respectively) may be corrected by the application of calibration data, stored in advance. Base-line detection eliminates drift and the disturbances caused by slit functions can be corrected by deconvolution.

- *Overlapping Peaks.* Positions of peaks can be found by differentiating the original spectrum as a function of time or wavelength. A stronger overlap asks for more consecutive differentiations in order to establish peak position. However, in practice the number of differentiations is limited because each differentiation enhances the noise. The peak surface is found by integration over time or wavelength. In order to obtain accurate peak surfaces the base line should be well defined.

- *A Base Line not Sufficiently Well Known.* The base line is found from those parts of a spectrum where no peaks are found. A base-line correction is only possible when such parts occur.

Another technique for separation of overlapping signals is a multiband resolution by means of curve fitting. Here it is known beforehand that the measured data follow a mathematical relationship

$$y = f(x) \tag{4.80}$$

and overlapping signals are additive. This implies for n signals

$$y(x) = \sum_{i=1}^{n} f_i(x) \tag{4.81}$$

The functions f_i are defined by a number of parameters such as peak position on the wavelength scale, peak intensity on the y-scale, half width at half height of the peak, intensity relations between groups of peaks, and so on. Each peak is described by its own function. In spectroscopy symmetrical peaks are mostly described by (FR 69) the Lorentz function,

$$Y_L = \frac{A_L}{1 + B_L^2 (x - x_{max})^2} \tag{4.82}$$

by the Gauss function,

$$Y_G = A_G \exp\left[-B_G^2 (x - x_{max})^2 \right] \tag{4.83}$$

or by linear combinations of these functions. A_L and A_G are found from intensity and/or normalization conditions such as $\int_{-\infty}^{+\infty} y\, dx = 1$. B represents a relationship with width at half height W: $B_L = 2/W$; $B_G = 2\sqrt{\ln 2}/W$.

Nonsymmetric peaks are approximated by means of empirically adapted standard digital functions. The estimated line shape can be approximated as accurately as desired by enhancement of the number of points in the digital function.

4.4.1 Curve Fitting

Curve fitting aims for a mathematical description of a series of measured data by a model with adaptable parameters. In general, the following procedure is applied: A mathematical model is chosen on the basis of empirical knowledge on the system under investigation (e.g., Lorentzian line shape or Gaussian line shape). The adaptable parameters in the model such as line width, peak position, and peak height are calculated in a manner that makes the sum of the squares of the deviations between calculated and measured data points as small as possible (FR 66, VA 75,

SE 79, SO 65). The number of parameters to be found equals the number of peaks multiplied by the number of parameters per peak. Suppose the analytical function that is going to represent the measured data equals (HY 70)

$$y = f(\alpha_1, \alpha_2, \alpha_3, \ldots, \alpha_n, x) \qquad (4.84)$$

where $\alpha_1, \ldots, \alpha_n$ are parameters to be determined such as line width and peak position. It is assumed that the number of measured data m is greater than the number of independent parameters n. The residual sum of squares R that represents the differences between the calculated values and the measured data Y_i is defined by

$$R = \sum_{i=1}^{m} \left[y_i - f(\alpha_1, \alpha_2, \ldots, \alpha_n, x) \right]^2 \qquad (4.85)$$

and has to be minimized with respect to each α_i. The condition

$$\frac{\partial R}{\partial \alpha_1} = \frac{\partial R}{\partial \alpha_2} \cdots \frac{\partial R}{\partial \alpha_n} = 0 \qquad (4.86)$$

yields n equations of the form

$$-2\left[y_i - f(\alpha_1, \alpha_2, \ldots, \alpha_n, x_i) \right] \frac{\partial f_i}{\partial \alpha_j} = 0 \quad (j = 1, 2, \ldots, n) \qquad (4.87)$$

$f(\alpha_1, \alpha_2, \ldots, \alpha_n, x_i)$ is now expanded in a Taylor series about a trial set of values $\alpha_j^0 (j = 1, 2, \ldots, n)$:

$$f(\alpha_1, \alpha_2, \ldots, \alpha_n, x_i) = y_i^0 + \left(\frac{\partial f}{\partial \alpha_1} \right) \Delta\alpha_1 + \left(\frac{\partial f}{\partial \alpha_2} \right) \Delta\alpha_2$$

$$+ \cdots + \left(\frac{\partial f}{\partial \alpha_n} \right) \Delta\alpha_n + \text{higher order terms} \qquad (4.88)$$

$y_i^0 = f(\alpha_1^0, \alpha_2^0, \ldots, \alpha_n^0, x_i)$ are the calculated values using the trial set of parameters, the starting values. A series expansion (eq. 4.88) with terms higher than the first-order neglected, is substituted in eq. 4.87, and with some rearrangement the following set of n linear simultaneous equations is obtained:

$$\sum_{i=1}^{m}\left(\frac{\partial f_i}{\partial \alpha_1}\right)^2 \Delta\alpha_1 + \sum_{i=1}^{m}\left(\frac{\partial f_i}{\partial \alpha_1}\right)\left(\frac{\partial f_i}{\partial \alpha_2}\right)\Delta\alpha_2$$

$$+\cdots+\sum_{i=1}^{m}\left(\frac{\partial f_i}{\partial \alpha_1}\right)\left(\frac{\partial f_i}{\partial \alpha_n}\right)\Delta\alpha_n = \sum_{i=1}^{m}\left(\frac{\partial f_i}{\partial \alpha_1}\right)(y_i - y_1^0)$$

$$\sum_{i=1}^{m}\left(\frac{\partial f_i}{\partial \alpha_1}\right)\left(\frac{\partial f_i}{\partial \alpha_2}\right)\Delta\alpha_1 + \sum_{i=1}^{m}\left(\frac{\partial f_i}{\partial \alpha_2}\right)^2 \Delta\alpha_2$$

$$+\cdots+\sum_{i=1}^{m}\left(\frac{\partial f_i}{\partial \alpha_2}\right)\left(\frac{\partial f_i}{\partial \alpha_n}\right)\Delta\alpha_n = \sum_{i=1}^{m}\left(\frac{\partial f_i}{\partial \alpha_2}\right)(y_i - y_i^0) \qquad (4.89)$$

$$\sum_{i=1}^{m}\left(\frac{\partial f_i}{\partial \alpha_1}\right)\left(\frac{\partial f_i}{\partial \alpha_n}\right)\Delta\alpha_1 + \cdots + \sum_{i=1}^{m}\left(\frac{\partial f_i}{\partial \alpha_n}\right)^2 \Delta\alpha_n = \sum_{i=1}^{m}\left(\frac{\partial f_i}{\partial \alpha_n}\right)(y_i - y_i^0)$$

The solution of the above normal equations yields $\Delta\alpha_j$, the first-order corrections to the trial set of values α_j^0. The new, improved set of values $(\alpha_j^0 + \Delta\alpha_j)$ is then used as a trial set, and the procedure is repeated until convergence is obtained. The iterative adjustments are necessary except for the simple linear case, where the first iteration yields the unique solution.

When a computer is available the calculations are most conveniently performed using the matrix representation of the set of equations 4.89.

This matrix representation of the residual sum of squares R is written as

$$R = \left[y - (y^0 + \Delta y)\right]^{\dagger}\left[y - (y^0 + \Delta y)\right] \qquad (4.90)$$

Here the brackets denote a column matrix and the † indicates the matrix transposition (rows and columns interchanged). The elements of the Δy matrix, Δy_i, represent the incremental change in f_i, introduced by (small) changes in the parameters α_j. It is assumed that none of the adjustments Δy_i is unduly large and that the adjustments are in the nature of refinement rather than gross changes. The relation between Δy and $\Delta\alpha$ is defined by

$$\Delta y = J \Delta\alpha \qquad (4.91)$$

J equals the Jacobian matrix and has elements

$$J_{i,j} = \frac{\partial f_i}{\partial \alpha_j} \qquad \begin{array}{l}(i=1,2,\ldots,m) \\ (j=1,2,\ldots,n)\end{array} \qquad (4.92)$$

It is seen that this relation makes the matrix description of R equivalent to that given by the Taylor expansion neglecting the higher-order terms.

Defining the error vector $\varepsilon = y - y^0$, whose elements ε_i are the difference between the experimental values y_i and the calculated value y_i^0 computed on the basis of the trial parameters α_j^0, and substituting eq. 4.92 in eq. 4.90 gives

$$R = [\varepsilon - J\Delta\alpha]^\dagger [\varepsilon - J\Delta\alpha] \qquad (4.93)$$

Upon differentiation and setting it equal to zero

$$\frac{\partial R}{\partial \Delta\alpha} = -2\varepsilon^\dagger J + 2\,\Delta\alpha\, J^\dagger J = 0 \qquad (4.94)$$

and transposing terms (note that the $J^\dagger J$ matrix is symmetric)

$$J^\dagger J \Delta\alpha = J^\dagger \varepsilon \qquad (4.95)$$

Expansion of this equation yields

$$\begin{vmatrix} J_{1,1} J_{2,1} \cdots J_{m,1} \\ J_{1,2} J_{2,2} \cdots J_{m,2} \\ J_{1,n} J_{2,n} \cdots J_{m,n} \end{vmatrix} \cdot \begin{vmatrix} J_{1,1} J_{1,2} & \cdots & J_{1,n} \\ J_{2,1} J_{2,2} & \cdots & J_{2,n} \\ J_{m,1} & & J_{m,n} \end{vmatrix} \cdot \begin{vmatrix} \Delta\alpha_1 \\ \Delta\alpha_2 \\ \Delta\alpha_n \end{vmatrix}$$

$$= \begin{vmatrix} J_{1,1} J_{2,1} & \cdots & J_{m,1} \\ J_{1,2} & & \\ J_{1,n} & & J_{m,n} \end{vmatrix} \cdot \begin{vmatrix} \varepsilon_1 \\ \varepsilon_1 \\ \varepsilon_m \end{vmatrix}$$

$$(4.96)$$

The solution of this equation for $\Delta\alpha$ is

$$\Delta\alpha = (J^\dagger J)^{-1} J^\dagger \varepsilon \qquad (4.97)$$

$(J^\dagger J)^{-1}$ is the inverse of $(J^\dagger J)$ and the elements of $(J^\dagger J)^{-1}$ can be obtained by the use of Cramer's rule when the matrix $J^\dagger J$ is nonsingular (the determinant is not zero).

It is convenient and sufficient in practice to construct an approximate Jacobian matrix for digital computation in the following way. A small increment ($\simeq 1\%$ of α_j^0) is given to the jth parameter α_j^0, the other parameters remaining unchanged. This set of parameters is then used to

calculate the m values of y_i'. The m values of $(y_i' - y_i^0)/\Delta \alpha_j$ then correspond to the jth column of the Jacobian matrix. The same increment is
then given to each of the other α_j's in turn, and y_i' is calculated for each set
of parameters. In this way all of the $m*n$ elements of the Jacobian matrix
are obtained.

The elements of the error vector ε are obtained by taking the differences
between the experimental values y_i and y_i^0 calculated with the trial set of
values α_j^0. After numerical substitution and matrix multiplication are
carried out a set of first-order corrections is obtained which determines a
new set of values for α_j. These values are used to calculate new values of
y_i^0 and a new Jacobian matrix. The cycle is repeated until convergence to
α_j is obtained. Convergence means that the elements of $\Delta \alpha$ have become so
small that the corresponding changes Δy_i are negligible.

Often in practice the Jacobian matrix need not be calculated for every
iterative cycle. When the trial set of values is reasonably close to the
converging set of values—good starting values—a single Jacobian matrix
can be used during the iterative operation. The final calculation, however,
should include a calculation of the last Jacobian.

The next step is to estimate the standard error of the parameters α_j. The
variance–covariance matrix of α is related to the Jacobian matrix by

$$V(\alpha) = (J^\dagger J)^{-1} \sigma^2 \qquad (4.98)$$

where the elements of the matrix are

$$V(\alpha) \begin{vmatrix} \sigma_1^2 & \sigma_1\sigma_2 & \cdots & \sigma_1\sigma_m \\ \sigma_2\sigma_1 & & & \\ \vdots & & & \\ \sigma_m\sigma_1 & & & \sigma_m^2 \end{vmatrix} \qquad (4.99)$$

The variance (square of the standard deviation) is the diagonal element in
$V(\alpha)$ and the covariance is the non-diagonal element. σ^2 is the variance of
the parent distribution and is estimated from the residual mean square s^2:

$$s^2 = \frac{1}{m-n} \sum_{i=1}^{m} \left[y_i - f(\alpha_1, \alpha_2, \ldots, \alpha_n, x_i) \right]^2 \qquad (4.100)$$

Now the α-parameters in f are the final set of the calculated values.

In eq 4.85 the residual sum of squares R is defined on an absolute basis. It is sometimes desirable, however, to fit the data on a percentage basis. This is particularly true when the magnitudes of the α_j values are spread over a wide range. Using W_i as a weighing factor, R can be redefined as

$$R = \sum_{i=1}^{m} W_i [\, y_i - f(\alpha_1, \alpha_2, \ldots, \alpha_n, x_i)]^2 \qquad (4.101)$$

and, when a similar derivation is carried out as above, $\Delta\alpha$ becomes

$$\Delta\alpha = (J^\dagger W J)^{-1} J^\dagger W \varepsilon \qquad (4.102)$$

where W is an $m * m$ diagonal matrix whose elements are $1/y_i^2$.

The iterative least-squares method of problem solution is feasible when computer facilities are at hand. However, satisfactory results are not always obtained. One of the main causes of difficulties is the convergence of the iteration procedure. This is illustrated in the three diagrams of Figure 4.22. In part a the procedure leads (slowly) to a correct answer (the dashed line); in part b the correct answer is not obtained, even after a great

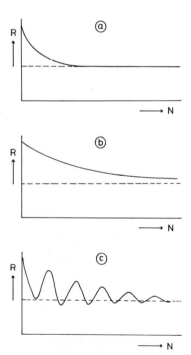

Figure 4.22. Various results of an iterative method of calculation.

number of iteration steps. Figure 4.22c depicts the situation where the final solution oscillates around the correct answer. The result depends on the number of the iteration step at which the calculation is disrupted.

An estimation of the correct answer may be obtained from averaging the results of the last few iteration steps; however, one cannot be sure that this is the correct answer.

The iterative least-squares method of problem solution is applied to situations where a number of equations—data—is available that exceeds the number of unknown parameters; for example, when a UV absorption spectrum of a mixture of two components is given as well as the individual spectrum of each of the components and the concentration ratio in the mixture is unknown, two extinction measurements would suffice to produce an answer. However, inclusion of all extinction measurements and a least-squares calculation enhances the accuracy and reliability of the answer.

Properties of curve fitting by least-squares procedures may be summarized as follows:

- All data must be stored in a data set before the calculations can be started.
- The base-line adaptation should be taken into account (WE 69).
- Partial derivatives with respect to all adaptable parameters should be calculated at each position of the spectrum. Sometimes this calculation must be repeated at each iteration cycle.
- Matrix manipulations are necessary.
- The starting parameters should be estimated. The possible dependence of the solution on the starting parameters must always be tested.
- Convergence may be accelerated by properly diminishing the parameter variations.
- The whole procedure may be a failure if an inappropriate mathematical model is selected.

In spectrum simulations the following limitations are set:

- The mathematical model for the line shape should be correct.
- Eventually a standardized digital line shape function may be applied; however, all signals must have the same line shape.

In situations where a spectrum contains Lorentzian and Gaussian lines it is not possible to simulate the entire spectrum with one line shape function.

In chromatographic spectra it is sometimes seen that high retention signals "tail" more than low retention signals. Here, too, one single line shape function cannot simulate all signals.

- The number of signals must be known before the iterative least-squares method is started. This knowledge may result from a theoretical consideration or from a test procedure that correlates the number of peaks in the model with the reduction of the average standard deviation between model and measurements.
- Even situations where the above conditions are fulfilled may not be uniquely solvable because of serious interaction between two or more of the parameters. Such situations are met with overlapping signals.

A new approach is the use of Kalman filters (Section 4.3.2), treated in SE 79 and PO 79. The problems encountered with spectrum simulations are dealt with in the references. (WE 69, PI 67, ST 62).

4.5 DATA HANDLING

In an analytical laboratory sometimes the relation between two measured quantities on a subject or a series of subjects is investigated. Two important questions that arise are, (1) What is the accuracy of the measurements and, related to this, (2) how accurate is the relation between the measured quantities? The problems related to these questions may be attacked by means of statistical techniques such as autocorrelation techniques and covariance calculations.

4.5.1 Covariance

In order to do an accurate measurement, it can be repeated a number of times, keeping the conditions of the measuring system as constant as possible. The result of such a series of measurements should be independent of time when the measuring system and the measuring quantities pertain to stationary states. In statistics the average result of such a series of measurements is sometimes referred to as expectation value $\varepsilon(x)$ or expectation of x:

$$\varepsilon(x) = \int_{+\infty}^{-\infty} xf(x)\,dx \qquad (4.103)$$

or

$$\varepsilon(x) = \sum_{-\infty}^{+\infty} xp(x) \qquad (4.104)$$

The first equation refers to a situation where x is described by a continuous probability distribution function $f(x)$, whereas the second equation is applied for discrete values of x occurring with a probability $p(x)$. The latter situation is met when a perfect die is thrown; here $p(x) = \frac{1}{6}$ for all values 1 up to and including 6 (for $x =$ number of dots) and $p(x) = 0$ for all other numbers. Here the expectation value $\varepsilon(x)$ equals

$$\varepsilon(x) = 1*\tfrac{1}{6} + 2*\tfrac{1}{6} + 3*\tfrac{1}{6} + 4*\tfrac{1}{6} + 5*\tfrac{1}{6} + 6*\tfrac{1}{6}$$

$$= 3.5$$

The definition of the expectation value of x can be extended to the expectation value of a function of x, $g(x)$ being a function of which x has an expectation value

$$\varepsilon[g(x)] = \sum_{-\infty}^{+\infty} g(x)p(x) \quad \text{for discrete } x\text{-values} \qquad (4.105)$$

and

$$\varepsilon[g(x)] = \int_{-\infty}^{+\infty} g(x)f(x)\,dx \quad \text{for continuous } x\text{-values} \qquad (4.106)$$

For $g(x) = ax$, a representing a constant, this results in

$$\varepsilon[g(x)] = \varepsilon[(ax)] = \sum_{-\infty}^{+\infty} axp(x) = a\sum_{-\infty}^{+\infty} xp(x) \qquad (4.107)$$

$$= a\varepsilon[x]$$

(the last equation also holds for the continuous probability distribution function).

Apart from the expectation value or mean value of x one may also be interested in the spread of x-values over the series of measurement in order to gain insight into the reproducibility of the measurements. The statistical expression applicable here is the variance of x:

$$\text{var}(x) = \varepsilon\{[x - \varepsilon(x)]^2\} \qquad (4.108)$$

$$= \sum_{-\infty}^{+\infty} [x - \varepsilon(x)]^2 p(x)$$

or in the situation with a continuous probability distribution of x,

$$\text{var}(x) = \int_{-\infty}^{+\infty} f(x)[x - \varepsilon(x)]^2 \, dx \tag{4.109}$$

Another expression used is the standard deviation $\sigma(x)$:

$$\sigma_x = [\text{var}(x)]^{1/2} \tag{4.110}$$

Again selecting the situation $\varepsilon(ax)$ one may derive

$$\text{var}(ax) = a^2 \text{var}(x) \tag{4.111}$$

and

$$\sigma(ax) = |a| \sigma_x \tag{4.112}$$

where $|a|$ is the absolute value of a.

Other useful equations that are easily derived are for $z = x + y$:

$$\text{var}(z) = \text{var}(x) + \text{var}(y) + 2\text{cov}(x, y) \tag{4.113}$$

Here the covariance between x and y is called $\text{cov}(x, y)$

$$\text{cov}(x, y) = \varepsilon\{[x - \varepsilon(x)][y - \varepsilon(y)]\} \tag{4.114}$$

In situations where the x-values are represented by a Gaussian probability distribution, $\varepsilon(x)$ gives the position of the maximum; $\text{var}(x)$ is related to the spread of x around the most probable value. $\text{Cov}(x, y)$ is related to the relationship between x and y values. Its meaning can be inferred from so-called contour diagrams. In Figure 4.23a high values of x are correlated with low values of y. This situation is described by a high negative covariance between x and y. In Figure 4.23b high x-values correspond with high y-values; here the $\text{cov}(x, y)$ is high and positive. In Figure 4.23c there is no correlation between x and y; thus $\text{cov}(x, y) = 0$. Figure 4.23d depicts a situation with a positive $\text{cov}(x, y)$; however, its magnitude is considerably lower than that of Figure 4.23b.

The magnitude of the covariance can be standardized. This standardized covariance is called the correlation coefficient $\rho(x, y)$:

$$\rho(x, y) = \frac{\text{cov}(x, y)}{\sigma(x)\sigma(y)} \tag{4.115}$$

in which $\sigma(x) = [\text{var}(x)]^{1/2}$. A special situation is met when $y = ax$ $(a \geqslant 0)$.

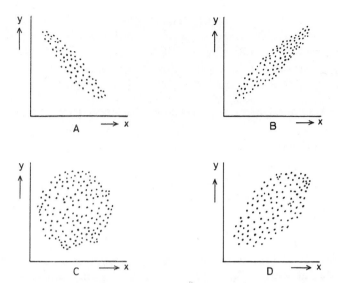

Figure 4.23. Correlation between x and y values. (a) Negative correlation with low spread; (b) positive correlation with low spread; (c) no significant correlation; (d) positive correlation with high spread (low significance).

Here

$$\text{var}(x) = a^2 \text{var}(x) \qquad (4.116)$$

$$\text{cov}(x, y) = \varepsilon\{[x - \varepsilon(x)][y - \varepsilon(y)]\}$$

$$= \varepsilon\{[x - \varepsilon(x)][ax - a\varepsilon(x)]\}$$

$$= a\,\text{var}(x) \qquad (4.117)$$

Thus

$$\rho(x, y) = \frac{a\,\text{var}(x)}{\sigma(x)\sigma(y)} = \frac{a\,\text{var}(x)}{\sigma(x)a\sigma(x)} = 1 \qquad (a \geqslant 0) \qquad (4.118)$$

In situations where the value found for x fixes completely the value to be expected for y, $\rho(x, y) = \pm 1$. Where no relationship exists $\text{cov}(x, y) = 0$ and thus $\rho(x, y) = 0$.

In the situations considered so far x as well as y pertain to stationary states. This changes when time-dependent processes are considered. Here the measured values x vary with time. The collected values x over a period of time are called a time series.

4.5.2   Time Series

A time series is a collection of measurements on the same variable quantity, measured as a function of time. Continuous time series are found in situations where the variable is continuously measured and recorded. (e.g., temperature, pressure). In analytical chemical laboratories many types of measurements are performed at discrete time intervals which produce discrete time series. The conversion of a continuous time series into a discrete one is achieved only by two methods:

1. By sampling of a continuous time series.
2. By accumulation over a given period of time (e.g, collection of rainwater).

From here on only discrete time series are discussed. In order to discuss the behavior of a time series the following notation is used: observations are made at series $T_0, T_1, T_2, T_3, \ldots, T_n$, giving observed values $Z(T_0), Z(T_1), \ldots, Z(T_n)$. In situations where the time interval between two successive measurements is a constant h the notation may conveniently be changed into

$$\text{measuring time } T_0, T_0+h, T_0+2h, \ldots, T_0+nh$$

$$\text{measurement } Z_0, Z_1, Z_2, \ldots, Z_n$$

For a starting time $T_0 = 0$ and h equal to the unit of time, Z_t becomes the measurement at time t. In practice two types of time series may be distinguished (BO 70):

- Deterministic time series in which the future is completely determined, such as

$$Z_t = \cos(2\pi f t) \tag{4.119}$$

- statistical time series, in which the values expected may be forecasted in terms of a probability distribution function.

A statistical phenomenon that develops in time according to probability laws is called a stochastic process (Figure 4.24). In order to analyze, that is, describe, a statistical time series it is considered to be a particular realization of a stochastic process.

A stationary process is a special type of stochastic process: it is a process in statistical equilibrium which means that the properties of the process do

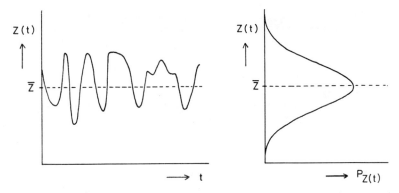

Figure 4.24. Example of a stochastic process $Z(t)$ as a function of time with its probability distribution function $P_{Z(t)}$.

not depend on the starting time of the observation period. Such a process can be subdivided into parts so that all may start at time T_0. This subdivision is allowed, provided a minimal realization length is obtained or surpassed. Such a signal is called an ergodic signal. Because the process is statistically determined, each time t provides a probability distribution for Z. This implies that a time series of M observations should be represented by an M-dimensional vector $(Z_1, Z_2, \ldots, Z_m)$ with a probability distribution $P(Z_1, Z_2, \ldots, Z_m)$. Given a stationary process, the condition holds that m measurements $Z_{t_1}, Z_{t_2}, \ldots, Z_{t_m}$ measured at time $t_1, t_2, \ldots, t_m$ are the same as the measurements $Z_{t_1+\tau}, Z_{t_2+\tau}, \ldots, Z_{t_m+\tau}$ measured at $t_1+\tau, t_2+\tau$. Two important properties of stationary processes are the following:

- The mean value μ is a constant given by

$$\mu = \varepsilon[Z(t)] = \int_{-\infty}^{+\infty} ZP(z)\, dz \qquad (4.120)$$

- The variance of Z is a constant

$$\sigma_z^2 = \varepsilon\left[(Z(t)-\mu)^2\right] = \int_{-\infty}^{+\infty} (z-\mu)^2 P(z)\, dz \qquad (4.121)$$

In the special situation $m=1$ the probability distribution $P(Z_t)$ is the same for all t-values and can be denoted as $P(z)$.

$P(z)$ can be found from a histogram of measurements on time series $Z(1), Z(2), Z(3), \ldots, Z(n)$ because $P(z)$ is a constant for all values of t. The average value μ from a stationary process is found from the average value of the time series

$$\bar{Z} = \frac{1}{n} \sum_{t=1}^{n} Z(t) \qquad (4.122)$$

and the variance

$$\sigma^2 = \frac{1}{(n-1)} \sum_{t=1}^{n} (z_t - \bar{Z})^2 \qquad (4.123)$$

(for large values of n, $n-1$ approaches to n).

For a stationary process the probability for occurrence for a pair of measurements $P(Z_{t_1}, Z_{t_2})$ is a constant for all times t_1 and t_2 provided the time interval between t_1 and t_2 is a constant. This is illustrated in Figure 4.25.

It is seen that the correlation between $Z(t+1)$ and $Z(t)$ is higher than that between $Z(t+5)$ and $Z(t)$. In general the correlation decreases upon an increase of τ.

Autocorrelation

The covariance between $Z(t)$ and $Z(t+\tau)$ of a time series can be calculated and is called the autocovariance γ_τ—*auto* because one stays within the time series—:

$$\gamma_\tau = \text{cov}[Z_t, Z_{t+\tau}]$$
$$= \varepsilon[(Z_t - \mu)(Z_{t+\tau} - \mu)] \qquad (4.124)$$

The collection of γ-values for various values of τ normalized to the

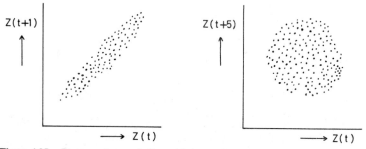

Figure 4.25. Decrease in correlation with increasing time span a time series $Z(t)$.

variance is called the autocorrelation function ρ_τ.

$$\rho_\tau = \frac{\varepsilon\left[(Z_t - \mu)(Z_{t+\tau} - \mu)\right]}{\left\{\varepsilon\left[(Z_t - \mu)^2\right]\varepsilon\left[(Z_{t+\tau} - \mu)^2\right]\right\}^{1/2}}$$

$$= \frac{\varepsilon\left[(Z_t - \mu)(Z_{t+\tau} - \mu)\right]}{\sigma_z^2} \qquad (4.125)$$

Because $\varepsilon[(Z_t - \mu)^2] = \varepsilon[(Z_{t+\tau} - \mu)^2] = \gamma_0$ for a stationary process,

$$\rho_\tau = \frac{\gamma_\tau}{\gamma_0} \qquad (4.126)$$

implying

$$\rho_0 = \frac{\gamma_0}{\gamma_0} = 1 \qquad (4.127)$$

Another name for γ_τ is the autocovariance function $\psi_{zz}(\tau)$, and for ρ_τ the autocorrelation function $\varphi_{zz}(\tau)$; it has no dimension.

$$\gamma_\tau = \rho_\tau \sigma_z^2 \qquad (4.128)$$

and

$$\psi_{zz}(\tau) = \varphi_{zz}(\tau)\sigma_z^2 \qquad (4.129)$$

Because τ is between $-\infty$ and $+\infty$ and the function is symmetric around $\tau = 0$, in general only that part of the function between 0 and $+\infty$ is represented. The time series given in Figure 4.26 yields the autocorrelation function. The autocorrelation function for $\tau = 0$ equals 1.0; a higher value of τ results in a lower value of $\varphi_{zz}(\tau)$. For the noncorrelated measurements the autocorrelation function sharply drops to 0 for all $\tau > 0$. Such an autocorrelation function also results from a series of random numbers, that is, white noise. In practical situations correlated time series frequently produce an autocorrelation function that may be represented by an exponential function

$$\varphi_{zz}(\tau) = e^{-\tau/T_P} \qquad (4.130)$$

Here T_P is called the time constant of the process. It is found from the τ-value at $\varphi_{zz}(\tau) = 0.37$, because $\ln 0.37 = -1 = -\tau/T_P$.

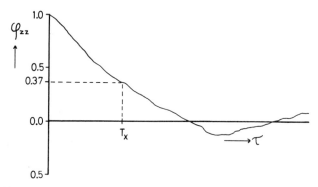

Figure 4.26. Autocorrelation function ϕ_{zz} of a time series $Z(t)$ as a function of time.

A nonstationary process yields $\varphi_{zz}(\tau)$-values that do not tend to zero for high τ-values. A measured signal that includes drift is always autocorrelated and produces a constant value for $\varphi_{zz}(\tau)$. A periodic function such as a sine function yields an autocorrelation function with the same periodicity and shape. A series of measurements suffering from many random high frequent disturbances can produce a very satisfactory autocorrelation function. The calculation of $\varphi_{zz}(\tau)$ may be useful in situations where a series of process values is hidden by sources of noise. Because white noise produces a zero contribution to the autocorrelation function for $\tau \neq 0$, the influence of random disturbances on the autocorrelation function is eliminated and the calculation of $\varphi_{zz}(\tau)$ offers a method of eliminating noise from a time series of measurements.

A stationary time series is characterized by the mean value μ and the autocovariance $\psi_{zz}(\tau)$ or alternatively by μ, the variance σ_z^2, and $\varphi_{zz}(\tau)$, the autocorrelation function.

A signal containing sufficient information for a process description is called an ergodic signal. The autocorrelation function of a stationary process has a limiting value zero for $\tau \to \infty$.

$$\lim_{\tau \to \infty} \varphi_{zz}(\tau) = 0 \tag{4.131}$$

and thus a signal with a zero-limiting autocorrelation function for $\tau \to \infty$ pertains to a stationary process and is an ergodic signal. This implies that a finite time series of measurements suffices for the process description.

Calculation of Autocorrelation Functions.   For a given time series $t_1, t_2, \ldots, t_n$ with $n \gg \tau$, $n - \tau \approx n$. The autocovariance equals

$$\psi_{zz}(\tau) = \frac{1}{n-\tau} \sum_{t=1}^{n-\tau} \left[Z(t) - \bar{Z}(t) \right] \left[Z(t+\tau) - \bar{Z}(t+\tau) \right] \tag{4.132}$$

For a stationary process

$$\bar{Z}(t) = \bar{Z}(t+\tau) = \bar{Z} \qquad (4.133)$$

thus

$$\psi_{zz}(\tau) = \frac{1}{n-\tau} \sum_{t=1}^{n-\tau} \left[Z(t)Z(t+\tau) + \bar{Z}^2 - Z(t)\bar{Z} - Z(t+\tau)\bar{Z} \right]$$

$$= \frac{1}{n-\tau} \sum_{t=1}^{n-\tau} \left[Z(t)Z(t+\tau) \right] - \bar{Z}^2 \qquad (4.134)$$

The variance equals $\psi_{zz}(\tau)$ for $\tau = 0$:

$$\sigma_z^2 = \frac{1}{n} \sum_{t=1}^{n} \left[Z(t) \right]^2 - \left[\frac{\sum_{t=1}^{n} Z(t)}{n} \right]^2 \qquad (4.135)$$

and the autocorrelation function

$$\varphi_{zz}(\tau) = \frac{\psi_{zz}(\tau)}{\sigma_z^2} \qquad (4.136)$$

Autocorrelation Function as Predicting Function. In most practical situations the autocorrelation function $\varphi_{zz}(\tau)$ of a time series $Z(\tau)$ equals

$$\varphi_{zz}(\tau) = e^{-\tau/T_P} \qquad (4.137)$$

This may be used to predict a process value for time t, based on the process value $Z(0)$ at time 0 (MU 78):

$$Z(t) = Z(0)e^{-\tau/T_P} \qquad (4.138)$$

where $Z(t)$ = predicted process value for time t
  $Z(0)$ = measured process value at time 0
  t = time span for the prediction
  T_p = time constant of the process $Z(\tau)$
The accuracy of the prediction $Z(t)$ decreases with an increased t-value. For limiting t-values toward ∞, a prediction of all values is possible within the limit of standard deviation and a probability corresponding to the Gaussian distribution.

Thus the autocorrelation function yields information on the transient behavior of a time series. An appropriately designed scheme for the analysis of variance (Section 4.6) can in principle give the same information. A relationship between the two techniques, with regard to their ability to verify and quantify a time constant of a first-order autoregressive stochastic stationary process, is described in the literature (LI 78).

Cross-Correlation

The autocorrelation function correlates a signal, for example, a series of measurements on a process, with itself. Another possibility arises when two different time series are intercorrelated, for instance, the input time series $x(t)$ and the output series $y(t)$ of a process at a time τ after the input measurement (Figure 4.27).

The autocovariance of $x(t)$, ψ_{xx}, and the cross-covariance function of $x(t)$ and $y(t)$, $\psi_{x,y}$, can be represented as shown in Figure 4.28. The mathematical expression for the cross-covariance function of x with y reads

$$\psi_{x,y}(\tau) = \varepsilon(\{x(t) - \varepsilon[x(t)]\}\{y(t+\tau) - \varepsilon[y(t+\tau)]\}) \quad (4.139)$$

It is seen that $\psi_{x,y}(\tau)$ for $\tau = 0$ is smaller than $\psi_{x,x}(\tau)$ for $\tau = 0$. An increase in τ results in an increased value for $\psi_{x,y}(\tau)$ until a maximal value is obtained; afterward the value of $\psi_{x,y}(\tau)$ decreases and tends toward 0. It is said that there is a time lag between input signal $x(t)$ and output signal

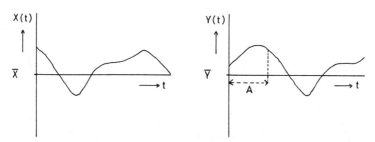

Figure 4.27. Input and output signals of a process. The output signal $y(t)$ has a delay A with respect to the input signal $x(t)$.

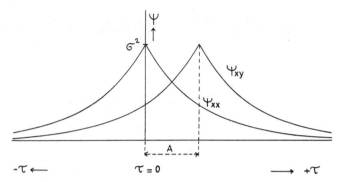

Figure 4.28. Autocovariance function ψ_{xx} and cross-covariance function $\psi_{x,y}$ of the signals from Figure 4.27. σ^2 represents the variance of the input signal $x(t)$; A is the delay between input $x(t)$ and output $y(t)$.

$y(t)$. In contrast to the autocovariance function the cross-covariance function is mostly not symmetric with respect to $\tau=0$. The position of the maximum cross-covariance is given by the time lag between $x(t)$ and $y(t)$:

A positive time lag means $x(t)$ is advanced with respect to $y(t)$.

A negative time lag means $x(t)$ is retarded with respect to $y(t)$.

A zero time lag means $x(t)$ and $y(t)$ are in phase.

Another property of $\psi_{x,y}(\tau)$ reads

$$\psi_{x,y}(\tau) \leqslant \left[\, \psi_{xx}(0) * \psi_{yy}(0)\,\right]^{1/2} \tag{4.140}$$

The covariance of the sum of independent signals equals the sum of the covariance of the individual signals

$$\psi_{x+y}(\tau) = \psi_x(\tau) + \psi_y(\tau) \tag{4.141}$$

The importance of cross-correlation function $\varphi_{x,y}(\tau)$ is found in its ability to describe a process. An input signal $x(t)$ corresponding with an autocorrelation function $\varphi_{xx}(\tau)$ determines an output signal $y(t)$ found from $\varphi_{x,y}(\tau)$.

Another procedure for identifying a process is the impulse response method. Here the response of a process to a well-known input pulse is characterized. An advantage of correlation functions over impulse response measurements is that, as well as autocorrelation, cross-correlation can be calculated from measurements available from the history of a process. In order to apply impulse response methods, the normal behavior of a process has to be disturbed. Apart from this it is sometimes rather difficult to produce a pulse disturbance on the input signal of a process.

4.6 ANALYSIS OF VARIANCE

The analysis of variance is a statistical technique for subdividing the total variance of a set of observations into partial variances that are attributed to well-defined sources. For example, when the quality of a basic material is given by the quantity of a certain chemical compound in the material one may measure the quality by analyzing a sample of the material. The analytical data obtained from various samples of the same material generally are distributed around an average value.

This distribution of results may be caused by the method of sampling, but also by the method of analysis. The aim of the analysis of variance is to attribute part of the variance to the method of sampling and another part to the method of analysis. An eventual improvement of the measurement of the quality of the material may result from a decrease in the variance of the distribution in the measurements. Here the cause of the most important source of variance can be found and its influence can be diminished.

4.6.1 Definitions

The variance s^2 of a set of n measured values x_i equals the arithmetic mean of the quadratic deviation of the measurements with respect to their average value $\bar{x}$:

$$s^2 = \frac{\sum_{i=1}^{n} (x_i - \bar{x})^2}{n-1} \tag{4.142}$$

with

$$\bar{x} = \frac{\sum_{i}^{n} x_i}{n} \tag{4.143}$$

In this equation n, the number of measurements in the set is limited. The same equations are used for the variance of an idealized set of measurements, where the population is increased to such an extent that the application of statistical theories is considered to be correct. The variance of such a set is called σ^2, the average value of the measurements μ, and the population M ($M \gg n$). In contrast to $\bar{x}$, μ is called the "true" mean value. Because of its magnitude, $M - 1$, the denominator in eq. 4.142 is converted into M.

Degrees of Freedom

When out of a set of three observations two observations are given, together with the average value of the set, the value of the third observation is fixed; for example, $x_1 = 2$, $x_2 = 4$, $\bar{x} = 5$ yields $x_3 = 9$. The number of degrees of freedom of this example is 2. Generally the number of degrees of freedom equals the number of independent data from a series of observations or, in other words, the total number of data diminished by the number of independent restrictions.

4.6.2   Conditions On Applicability

In the method of analysis of variance, a considerable number of observations are involved, collected according to a previously designed scheme. The observations are divided into groups resulting from various samples, all belonging to a given population. In order to compare the results from different samples a number of conditions are set:

1.   The population investigated should be at least approximately normally distributed, in order to allow calculation methods based upon the variances and the application of the F-test.

2.   All samples must be taken from the same population. Comparison of data resulting from different methods of analysis or different instruments in order to investigate the effect of another method or instrument of course involves data from different populations. Here the method of comparing variances looks the same but the conclusions to be drawn are quite different, namely, the preference of one method or instrument over another.

3.   The population on which the measurements are made should be constant in time. The measurements are made consecutively; therefore a time dependence of the population may result in a trend in the collected data and thus a wrong estimate of the variance can be the consequence.

4.   The sampling of the probes under investigation should be such that they are normally distributed because here too calculations of variances and F-tests are performed.

5.   All probes should have the same variance. This condition is crucial for the analysis of variance method. The equal variance in combination

with the equal mean value of all probes is called the null hypothesis of the analysis and, based on the F-test, the acceptance or rejection of this hypothesis provides the conclusion about the unequal contribution of variances by various sources. The design of a proper analysis of variance scheme allows at least two different methods for estimation of the common variance of the total population. When two of the methods provide different estimates of the mean value, and these differences are proved to be significant, it is concluded that the averages of the samples are different. The null hypothesis is rejected.

The method is illustrated by the following example. From a given population of objects k samples are selected, each containing n_k objects. The n_k objects of sample k are normally distributed with a variance σ_k^2 that equals σ_0^2. Here σ_0^2 is the variance of the parent population. The question to be asked is whether the mean values $\bar{x}_k$ of the samples k are equal or not. This question is formulated in the null hypothesis H_0:

$$\mu_i = \mu_0 \qquad (i = 1, 2, 3, \dots, k) \tag{4.144}$$

The analysis of the variance scheme designed to answer this question can be summarized as follows. Calculate for each sample j:

1. The number of observations made on sample j equals n_j; the observations themselves are $x_{i,j}$ $(i = 1, 2, \dots, n_j)$.
2. The average value of x found for sample j equals

$$\bar{x}_j = \frac{\sum\limits_{i=1}^{n_j} x_{i,j}}{n_j} \tag{4.145}$$

3. The variance of the observations made on sample j equals

$$s_j^2 = \frac{\sum\limits_{i=1}^{n_j} (x_{i,j} - \bar{x}_j)^2}{n_j - 1} \tag{4.146}$$

Thus the k calculated values s_j^2 all are estimates of the population variance σ_0^2 of the parent distribution; this is a condition set in advance in order to apply the analysis of variance method. This being so, all s_j^2-values may be combined in order to yield a better value for σ_0^2. This value equals

$$s_w^2 = \frac{1}{k} \frac{\sum\limits_{j=1}^{k} \sum\limits_{i=1}^{n_j} (x_{i,j} - \bar{x}_j)^2}{n_j - 1} \tag{4.147}$$

This better estimate of σ_0^2 is based on the variance estimates within the sample, and is called the within-group variance.

Now another way is designed to obtain an estimate of σ_0^2, starting from the mean values $\bar{x}_j$ of all samples k. Suppose these $\bar{x}_j$-values stem from a population with variance σ_0^2. The variance of the $\bar{x}_j$-values equals

$$s_{\bar{x}}^2 = \frac{1}{n_j}\sigma_0^2 \tag{4.148}$$

This variance is also found from

$$s_{\bar{x}}^2 = \frac{1}{k-1}\sum_{j=1}^{k}\frac{n_j}{n}(\bar{x}_j-\bar{x})^2 \tag{4.149}$$

where $\bar{n}=\sum_{j=1}^{k}(n_j/k)$ and in which

$$\bar{x} = \frac{\displaystyle\sum_{j=1}^{k}\sum_{i=1}^{n_j}x_{i,j}}{\displaystyle\sum_{j=1}^{k}n_j} \tag{4.150}$$

When it is assumed (eq. 4.149) that all samples k contain the same number of measurements, then $n_j = n$ for all values of j. For unequal values of n_j the terms $(\bar{x}_i-\bar{x})^2$ should be multiplied by a proper weighting factor $n_j/\bar{n}$ in order to equalize the contribution of all measurements $x_{i,j}$ to the variance.

This second estimate of σ_0^2, which is based on the variance between the mean values of the samples, is called the between variances s_b^2. Its value equals $ns_{\bar{x}}^2$ provided all samples contain the same number of measurements n.

The null hypothesis states that s_b^2 is an estimate of σ_0^2; s_w^2 is always an estimate of σ_0^2. If s_b^2 and s_w^2 are unequal, tested for significance by the F-test, this cannot be caused by unequal values for the variance, because the application of the analysis of variance scheme implies equal values for sample variances. Therefore the mean values $\bar{x}_k$ should be unequal, and the null hypothesis has to be rejected. Equal values found for s_b^2 and s_w^2 imply that the null hypothesis is correct and that s_b^2 and s_w^2 may be combined in order to obtain a good estimate s_0^2 for σ_0^2

$$s_0^2 = \frac{(n_1-1)s_w^2+(n_2-1)s_b^2}{n_1+n_2-2} \tag{4.151}$$

in which n_1-1 equals the number of degrees of freedom in s_w^2 that is, $k(n-1)$, and n_2-1 equals the number of degrees of freedom in s_b^2, that is, $k-1$. Thus

$$s_0^2 = \frac{\displaystyle\sum_{i=1}^{n}\sum_{j=1}^{k}(x_{i,j}-\bar{x}_j)^2+n\sum_{j=1}^{k}(\bar{x}_j-\bar{x})^2}{kn-1} \tag{4.152}$$

It is assumed that $n_j = n$ for all values $j = 1,\ldots,k$.

If H_0 has to be rejected s_b^2 is not a good estimate of σ_0^2 but s_w^2 is. Here s_b^2 is interpreted as the true variance of the parent population, increased by an amount $n\sigma_\mu^2$. σ_μ^2 equals the variance of the population of the mean values μ_i of the k samples, each with n numbers. The estimated value of σ_μ^2 equals s_μ^2, and is found from

$$s_\mu^2 = \frac{1}{n}\left(s_b^2 - s_w^2\right) \qquad (4.153)$$

Generally $s_\mu^2 \geqslant 0$, so $s_b^2 \geqslant s_w^2$. If $s_b^2 < s_w^2$ the difference between the two should not be significant. If the difference is significant, a mistake has been made in the calculations or in the measurements.

A nonzero contribution s_μ^2 to the variance between samples is interpreted as the introduction of uncertainty to the analytical results by the method of sampling. s_w^2 is the uncertainty in the analytical results caused by the method of analysis itself or by the distribution of x in the parent population.

The method can be extended to more variables by grouping all measurements in such a way that each time one parameter of a group is kept constant. The difference between the estimated variance of such a group and the best estimated value of σ_0^2 each time is tested on significance and if this difference is significant, attributed to uncertainty of the parameter under investigation.

Table 4.14 is an example of an analysis of variances scheme. When the measurements of Table 4.15 are used, Table 4.16 results.

4.6.3 Hierarchical Classification

The example presented above can be depicted as in Figure 4.29. The total variance s_0^2 of all analytical results equals the sum of the contributions of

TABLE 4.14   Analysis of Variance Scheme

Source of Variance	Sum of Squares	Number of Degrees of Freedom	Mean Square	Quantity Estimated
Between samples	$S_b = n \sum_{j=1}^{k} (\bar{x}_j - \bar{x})^2$	$\nu_1 = k - 1$	$s_b^2 = \dfrac{S_b}{k-1}$	$\sigma_0^2 + n\sigma_\mu^2$
Within samples	$S_w = \sum_{j=1}^{k} \sum_{i=1}^{n} (x_{i,j} - \bar{x}_j)^2$	$\nu_2 = k(n-1)$	$s_w^2 = \dfrac{S_w}{k(n-1)}$	σ_0^2
Total	$S_t = \sum_{j=1}^{k} \sum_{i=1}^{n} (x_{i,j} - \bar{x})^2$	$\nu_3 = nk - 1$		

TABLE 4.15 Numerical Example of an Analysis of Variance with Two Variables[a]

	Analytical Data of Five Analyses on Six Samples					
Sample No.	1	2	3	4	5	6
Analysis no.						
1	145	140	195	45	195	120
2	40	155	150	40	230	55
3	40	90	205	195	115	50
4	120	160	110	65	235	80
5	180	95	160	145	225	45
Average value of sample, $\bar{x}_j$	105	128	164	98	200	70

[a] The $x_{i,j}$ values, denoting the amount of a given chemical compound of a sample j, are listed. Each sample has been analyzed five times ($i = 1, 2, \ldots, 5$).

TABLE 4.16 Analysis of Variance Scheme, Numerical Example

Source of Variance	Sum of Squares	Number of Degrees of Freedom	Mean Square	Quantity Estimated
Between samples	$S_b = 56{,}357$	$\nu_1 = 5$	$s_b^2 = 11{,}272$	$\sigma_0^2 + 5\sigma_\mu^2$
Within samples	$S_w = 58{,}830$	$\nu_2 = 24$	$s_w^2 = 2{,}451$	σ_0^2
Total	$S_t = 115{,}187$	$\nu_3 = 29$		

Conclusions

$s_b^2 = 11{,}272 \quad s_w^2 = 2451 \rightarrow s_b^2 > s_w^2$

$F_{(\nu_1, \nu_2, \alpha)} = F_{(5, 24, 0.05)} = 2.62$ (See Fisher table 3.10)

$$F = \frac{s_b^2}{s_w^2} = 4.50$$

$F > F_{(\nu_1, \nu_2, \alpha)} \rightarrow s_b^2$ is significantly higher than s_w^2

H_0 rejected $s_\mu^2 = \dfrac{s_b^2 - s_w^2}{5} = 1764$

s_μ^2 is an estimate of σ_μ^2, the variance of the average analyses of the samples

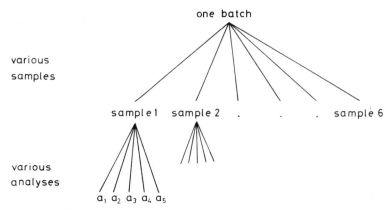

Figure 4.29.   Hierarchical classification in an analysis of variance scheme.

each level in the hierarchical one. This scheme may easily be extended to more levels, each with its own cause of errors.

4.6.4   Cross-Classification

This type of classification aims for classification of observations in a manner that is independent of all other causes of errors. As an example, in comparing three different analytical procedures in five different laboratories the variable parameters are the laboratories and the procedures. The laboratories and the procedures are completely independent and the results of the measurements can be grouped according to laboratories and according to analytical procedures. Discrepancies between laboratories can be found from the average analytical results of the various procedures, discrepancies between procedures from the average results of various laboratories. All data suffer from errors in the measurements or sampling

TABLE 4.17.   Analysis of Variance Scheme (Cross-Classification)

Laboratory	Procedure			
	1	2	...	m
1	$x_{1,1}$	$x_{1,2}$		$x_{1,m}$
2	$x_{2,1}$			
3			$x_{r,c}$	
4				
:				
:				
n	$x_{n,1}$			$x_{n,m}$

TABLE 4.18. Analysis of Variance Scheme for Cross-Classification[a]

Cause of Variation	Sum of Squares	Degrees of Freedom	Mean Square	Quantity Estimated
Between columns (procedure)	$A = n \sum_{j=1}^{m} (\bar{x}_j - \bar{x})^2$	$\nu_1 = m - 1$	A/ν_1	$\sigma_0^2 + n\sigma_m^2$
Between rows (laboratories)	$B = m \sum_{j=1}^{n} (\bar{x}_j - \bar{x})^2$	$\nu_2 = n - 1$	B/ν_2	$\sigma_0^2 + m\sigma_n^2$
Total	$C = \sum_{i=1}^{n} \sum_{j=1}^{m} (x_{i,j} - \bar{x})^2$	$\nu_t = n*m - 1$		

[a]m = number of procedures; n = number of laboratories; $x_{i,j}$ = measurement of lab i and method j.

which contribute to the standard deviation of the interlaboratory as well as the interprocedure σ^2. The cross-classification scheme can be extended to more than two variables.

The matrix describing the cross-classification scheme for n laboratories and m procedures is given in Table 4,17.

The analysis of variance scheme (DA 71, DA 72) is given in Table 4.18. In this scheme $C \neq A + B$ for most situations. This may be caused by analysis errors and mutual interaction effects between laboratories and procedures, that are not taken into account. Sometimes the value of $C - (A + B)$ is called the rest variance. When no interaction effects are expected between the two parameters the value of $C - A - B$ equals an estimate of σ_0^2.

4.7 PATTERN RECOGNITION

Pattern recognition may be described as a tool designed to aid the scientist in making an educated guess (DE 75) in the systematic analysis of multidimensional data when direct or statistical analysis is not feasible and calculation based on a physical model is impossible. This is a rather general statement; more specifically, pattern recognition finds and/or predicts a property of the objects that is not measurable but is known to be related to the measurements via some unknown relationships, given a set of objects and a list of measurements made on those objects (KW 72, KO 73, AN 72, DU 73).

In this description a property is the information (one hopes) implicit in the data one obtains, for example, the origin of some material (using elemental composition measurements) (KO 72, DU 75, DW 75), the quality of some material (using chemical and physical measurements) (KW 74), the structure of some compound (using spectral measurements) (KO 71, KO 74), or the chemical and/or biological activity of some compound (using molecular structure and composition information) (KA 74, RE 74). In this context properties can be considered to lie along a continuous scale ranging from discrete category to continuous property.

Discrete category data has as its property an *identifier* which may be arbitrarily assigned (material problems are often of this nature). The value of such a property is not intended to be a function of the measurements; it is assigned to permit identification of the discrete types or groups—categories—of objects. Continuous property data have a value that is assumed to be a function of the measurements. Real applications often fall between these extremes into what are termed continuous categories. Reactivity data taken as continuous values are often no more accurate than could be described by low, middle, and high. This range of possible property types would be of little importance except that the two extremes of discrete category and continuous property require different pattern recognition methods.

Generally, the use of discrete categories or continuous categories allows more, and more powerful, pattern recognition methods to be applied. This is intrinsic to the simpler "is it or is it not?" question that the categories pose, in comparison to the question of "how much" posed with continuous property. However, the arbitrary nature of the category identifier in a truly discrete category situation makes the analysis of that problem with continuous property methods meaningless. A reassignment of the category identifiers gives very different results. (An exception is the situation in which only two categories are present: switching the assignment merely interchanges the signs of the results.)

The approach based on pattern recognition algorithms often gives good results, but it cannot be considered universal, since in each particular case the computer has to recognize some new classes of compounds. As a rule, pattern recognition algorithms are used to solve only individual problems, without any real claim to versatility.

Computer learning aimed at the recognition of a new class of compounds cannot be carried out successfully unless the training data are adequate and representative. The narrow specialization of the identification systems incorporating pattern recognition algorithms is a major disadvantage of the approaches mentioned above (GR 77).

4.7.1   Data Collection and Presentation

The measured data on objects may result from various one-dimensional analytical procedures such as melting point, boiling point, or refraction measurement. Here the set of measurements is denoted as multisource. Measured data exclusively resulting from one two-dimensional method such as spectral analysis are named single-source. Here the two dimensions are spectral position and intensity. Both types of data may be used in pattern recognition procedures.

The collecting of multisource data is done by making tables, computer punch cards, or data records in an input file of a computer. Single-source data can automatically be collected by means of on-line dedicated computers belonging to the measuring apparatus.

When signal-to-noise ratio is a problem, single–source data may be sampled many times and coherently added. Modern spectroscopic techniques often have built-in facilities for such a procedure and the signal-to-noise improvement is visually controllable on an oscilloscope.

In order to apply pattern recognition, all measured data on one object are conveniently represented as a vector x of dimension n in a hypergeometric space R^n. Here n equals the dimensionality and the number of measured properties per object.

A spectrum of a compound can be represented as a number of frequencies n, each with an intensity I_n. The collection of n intensities provides one compound vector x_n. Comparison of two different spectra can be considered as measuring the distance between the end points of two vectors x_i and x_j of the same dimensionality in the representative space R^n.

Pattern recognition is applied in the process of comparing a large collection of representation points in order to detect clusters of points. Such a cluster is attributed to objects with like properties.

4.7.2   Preprocessing Techniques

Sometimes the measured data forming n-dimensional patterns in the hypergeometric space are subjected to transformation procedures before the actual pattern recognition techniques are applied (JU 75). These transformations serve the following purpose:

• The separation of various clusters of like vectors is improved in order to simplify the classification.

- The dimensionality of the pattern space R^n is reduced in order to economize on the computing time and to visualize the result.

One method of preprocessing that may be an advantage is autoscaling. This technique distributes the measured data in such a manner along the axis of the coordinate system that the average value of the coordinates along each of the axes equals zero and the variance equals one. This procedure aims for an equal resolution along each of the axis. The values of the coordinates in the pattern space R^n thus change by autoscaling. The jth coordinate value of pattern vector a becomes

$$x'_{aj} = \frac{x_{aj} - \bar{x}}{\sigma} \qquad (4.154)$$

in which x'_{aj} represents the autoscaled jth coordinate of pattern x_{aj} the original value, $\bar{x}_j = (1/N)\Sigma_{a=1}^{N} x_{aj}$ the average value of j for all N patterns and $\sigma_j^2 = \Sigma_{a=1}^{N}(x_{aj} - \bar{x}_j)^2/(N-1)$ the variance of the j-coordinate of all patterns.

4.7.3 Recognition without Supervision

Recognition here is the discovery of clusters of patterns in the n-dimensional hypergeometric space R^n. In situations where n equals 2 or 3 the technique for recognition that is applied most successfully is visual recognition because human visual perception is a very powerful method. When the dimensionality of the hyperspace exceeds 3, two ways are open:

- Dimension reduction.
- Mathematical recognition methods.

Dimension reduction may proceed in two different ways:

1. Linear combinations of features along the different axes in the n-dimensional space may be constructed in such a way that n features produce n orthogonal linear combinations. These orthogonal combinations are ordered according to decreasing variance. Because the variance of a particular feature—or feature combination—contains the information content of these features for the proper description of the patterns, the combination with the highest variance is the most significant one for the pattern description. In practical situations it sometimes occurs that the two or three topmost feature combinations—"eigenvectors"—encompass a considerable percentage of all available information (sum of "eigenvalues").

In such situations one may depict these two or three eigenvectors along three orthogonal axes in order to survey the entire problem without causing too much uncertainty by omitting the other eigenvectors. Hereafter one may rely on visual inspection. The mathematical procedure for the construction of eigenvectors and eigenvalues is known as the Karhunen–Loève transformation; it operates on the variance–covariance matrix of the various features (KO 73).

2.   In the nonlinear mapping procedure, all points in the R^n space are projected into the R^2-space under the condition that the mutual distance between each pair of points remains the same. When the directions along the two axes in the R^2-space are called x and y, this condition for two points a and b implies

$$D_{ab} = D_{ab}^* = \left[(x_a - x_b)^2 + (y_a - y_b)^2 \right]^{1/2} \qquad (4.155)$$

where D_{ab} equals the distance between a and b in R^n and D_{ab}^* in R^2. For many points these distances cannot all be transformed without making errors; therefore an error is defined according to

$$E = \frac{\sum\limits_{a>b} (D_{ab} - D_{ab}^*)^2}{D_{ab}} \qquad (4.156)$$

when the summation sums over all pairs of points and

$$D_{ab} = \left[\sum\limits_{j=1}^{n} (x_{a,j} - x_{b,j})^2 \right]^{1/2} \qquad (4.157)$$

This error E represents a nonlinear function of $2N$ unknowns (the x and y values of N points in the R^2-space). Minimalization of E with respect to all unknown produces the position of all points in R^2.

Mathematical recognition methods start from the position of all points in R^n. These positions are transformed into a distance matrix that lists all distances between all points or into an equality matrix G which is related to the distance matrix according to

$$G_{a,b} = 1 - \frac{D_{a,b}}{D_{\max}} \qquad (4.158)$$

where $D_{\max}$ equals the distance between the two points that are farthest apart. It is seen that these two points produce an equality 0 whereas two points at the same location give $G = 1$.

A particular method for cluster search, called hierarchical or Q-mode clustering, is based on the equality matrix. Here the pair of points with highest equality are sought and substituted for their center of gravity. This center of gravity gets an associated weight constant, denoting the number of points involved in the substitution. This process is repeated for the remaining points until one of the following occurs:

- A number of clusters—or gravity centers—is obtained equal to a fixed number set in advance.
- The highest remaining equality value is lower than a limiting value set in advance.
- The decrease in equality on repetition of the clustering procedure surpasses a limiting value.

It should be mentioned here that instead of Euclidean distance matrices— with an exponent value $\mu = 2$ in $[\Sigma(x_{aj} - x_{bj})^{\mu}]^{1/\mu}$—sometimes other exponents are used (e.g., $\mu = 1$). The general name for such distances is Minkovski distances.

The position of the patterns in R^2 can be visualized by various methods. A very useful one is that using the line printer of a modern computer. Here the patterns, denoted by their number or name, are printed in a position corresponding with the x and y values in R^2. Instead of a pattern number one may also use the class number in situations when supervised classification can be applied. A disadvantage can be that two or more patterns may be located at the same position. The computer program "Arthur" (DW 75) provides such a printing routine and a bookkeeping system that counts the number of multiple points for one position.

Plotting with xy-plotters is another useful method. These plotters also offer the possibility of constructing stereoscopic pictures in an R^3-space by making two pictures in R^2, rotated about 15 degrees and 15 cm apart, which can be inspected with a stereoscope in a manner that is well-known to crystallographers.

These graphs can be of help during the classification process. In specialized computing centers one may apply television screens and on-line rotation–transformation routines that allow rotation of a two-dimensional picture along three perpendicular axes. Here the advantage of a R^3 picture is obtained by rotations of the R^2 presentation on command of the observer until the best picture is seen.

4.7.4 Training and Classification

Up to now the recognition methods applied refer to situations in which it is not known whether or how many clusters exist. Sometimes the situation is not so bad, and the number of clusters or classes of compounds is known. Here pattern recognition methods with supervision may be applied that use the information on the number of classes given. Such procedures are called training and classification. In this context training pertains to the methods applied to classify the various clusters and it aims for the development of a strategy that can be successfully used when new unknown samples should be classified. Thus training is the development of decision operators using well-classified patterns aiming for correct classification of unknown patterns. Methods with supervision start with attributing weights to the various features, taking into account their value for class separation. This feature weighing is part of the preprocessing technique.

4.7.5 Feature Weighing

In a particular pattern recognition method, not all measured data—features—of an object are equally important for the description of the property associated with the patterns. In order to evaluate the individual importance of each feature, evaluation rules (weighing functions) should be applied to select the most important ones. Three different weighing procedures may be mentioned here.

Variance weighing is a procedure that measures the utility feature j has for the separation of category m from category n. The algorithm for the value of this utility $W_{j,m,n}(v)$ is

$$W_{j,m,n}(v) = \frac{\left[\overline{(x_m)_j^2} + \overline{(x_n)_j^2} - 2\overline{(x_m)_j}\, \overline{(x_n)_j} \right]}{(m2)_{m,j} + (m2)_{n,j}} \tag{4.159}$$

in which $(m2)_{m,j}$ equals the second moment of feature j in class m $[=(N_{jn}-1)\sigma_{m,j}^2/N_{jn}]$. The variance weight of feature j for all linear class separations $W(v)_j$ equals, for M classes,

$$W(v)_j = \left[\mathop{\pi}_{m=1}^{M-1} \mathop{\pi}_{n=m+1}^{M} W(v)_{j,m,n} \right]^{2/M(M-1)} \cdot \tag{4.160}$$

An increase in correlation between the feature j of category m and n here results in a decrease in the utility of this feature for the separation.

Another weighing is that according to Fisher. Here the weighing factor equals

$$W_{j,m,n}(F) = \frac{\left[\overline{(x_m)}_j - \overline{(x_n)}_j \right]^2}{N_m \sigma_{m,j}^2 + N_n \sigma_{n,j}^2} \qquad (4.161)$$

The population of a class is taken into account in this equation. The Fisher weight of feature j for all linear class separations $W(F)_j$ equals

$$W(F)_j = \frac{2 \left[\sum\limits_{m=1}^{M-1} \sum\limits_{n=m+1}^{M} W(F)_{j,m,n} \right]}{M(M-1)} \qquad (4.162)$$

A weighing factor that takes into account the value of feature j for separating property p_k from other properties is

$$W_j(p) = \frac{\left[\sum\limits_{k=1}^{N} (x_{j,k} - \bar{x}_j)(p_k - \bar{p}) \right]^2}{\sigma_j^2 (N-1)^2 \sigma_p^2} \qquad (4.163)$$

or, in another notation,

$$W_j(p) = \frac{\left[\sum\limits_{k=1}^{N} (x_{j,k} - \bar{x}_j)(p_k - \bar{p}) \right]^2}{\sum\limits_{k=1}^{N} (x_{j,k} - \bar{x}_j)^2 \sum\limits_{k=1}^{N} (p_k - p)^2} \qquad (4.164)$$

in which N denotes the number of patterns in the entire training set, $\bar{x}_j$ equals the average value of feature j, and $\bar{p}$ is the average value of a property. In a shorter notation, with $x_{j,k} - \bar{x}_j = x_{j,k}^1 (p_k - \bar{p}) = p_k^1$ and $\overline{ab} = \sum_{j=1}^{N} a_j b_j / N$,

$$W_j(p) = \left[\frac{N}{N-1} \cdot \frac{\overline{x_j^1 p^1}}{\sigma(x_j)\sigma(p)} \right]^2 = \frac{(\overline{x_j^1 p^1})^2}{m2(x_j)m2(p)} \qquad (4.165)$$

4.7.6 Recognition with Supervision

A conceptual simple method for recognition with supervision is that based on nearest neighbors. Here the class to which a particular compound belongs is estimated from the class of its nearest neighbors. "Near" is defined from the distance matrix and in case of more neighbors the majority wins. The number of neighbors taken into account is found from statistical considerations and depends on the acceptable risk for misclassification and the number of members in the training set.

Another method is the development of linear pattern classifiers by means of negative feedback procedures. In order to apply these methods a training set should be available encompassing all classes and with about equal population in all classes. The aim now is the construction of a number of planes that each subdivide all patterns into two groups. The decision planes thus developed should all pass through the origin of the set of coordinate axes. In order to reach this goal, the number of coordinates that describes the position of a pattern in the R^n-space is extended with one extra dimension (mostly chosen equal to 1). Now the orientation of a particular decision plane is fixed by a vector $\mathbf{w}$, perpendicular to this plane, at the origin. Multiplication of the vector $\mathbf{w}$ by the pattern vector $\mathbf{y}$ produces a scalar $s = \mathbf{w} \cdot \mathbf{y} = \mathbf{w}\mathbf{y}\cos\Theta$ in which Θ denotes the angle between $\mathbf{w}$ and $\mathbf{y}$. Depending on Θ, the cosine function is positive or negative. The sign of the s is used to develop the orientation of the decision plane: each member of the training set is multiplied by the decision vector $\mathbf{w}$ and each time a pattern is incorrectly classified the orientation of the decision plane and thus that of $\mathbf{w}$ is altered, for example, by a reflection operation in such a way that the distance between the pattern and the plane remains constant.

In situations where a linear separation between patterns is possible, the consecutive alterations of $\mathbf{w}$ should diminish and converge to a final value for $\mathbf{w}$. Sometimes, however, the patterns are not linearly separable and no convergence is achieved. Here the calculations, mostly performed with computer facilities, should be disrupted.

The procedure should be repeated for each set of two classes out of the training set. The set of decision vectors thus developed can be used to classify unknown patterns. The reliability of the process increases with an increase of the number of patterns in the training set (see Table 4.19).

Instead of negative feedback one may also apply a least-squares procedure to train the decision vectors. Here the difference between the actual s value s^* and the one calculated according to $\mathbf{w} \cdot \mathbf{y}$ is minimized for all y-values as a function of all $n+1$ elements of $\mathbf{w}$ (n = dimensionality of the problem = number of features).

$$s_i = \mathbf{w}\mathbf{y}_i = w_1 y_{i,1} + w_2 y_{i,2} + \cdots + w_{n+1} y_{n+1} \tag{4.166}$$

The error vector between s_i^* and s_i to be minimized,

$$R = \sum_{i=1}^{N} \{s_i - s_i^*\}^2 \tag{4.167}$$

**TABLE 4.19.   Mathematical Procedure for Training a
Decision Vector According to a Negative
Feedback Procedure**

1. Calculate $s = \mathbf{w} \cdot \mathbf{y}_a$: a sign wrong, $\mathbf{w} = \mathbf{w}_{old}$; $s = s_{old}$,
                                b sign right, take new y-value
2. Calculate $\mathbf{w}_{new} = \mathbf{w}_{old} + c \mathbf{y}_a$ with $c = \dfrac{-2 s_{old}}{\mathbf{y}_a \cdot \mathbf{y}_a}$
3. Calculate $s_{new} = \mathbf{w}_{new} \cdot \mathbf{y}_a$

$$= \mathbf{w}_{old} \mathbf{y}_a - \frac{2 s_{old} \cdot \mathbf{y}_a \cdot \mathbf{y}_a}{\mathbf{y}_a \mathbf{y}_a}$$

$$= + s_{old} - 2 s_{old}$$

$$= - s_{old}$$

4. Repeat step 1 with a new y-value
5. Stop the calculation when corrections are no longer required
    for all y-values involved

where N = number of patterns in the training set, yields $n + 1$ equations:

$$\frac{\partial R}{\partial w_k} = 2 \sum_{i=1}^{N} \left[\left(\sum_{j=1}^{n+1} w_j y_{i,j} - s_i^* \right) y_{i,k} \right] = 0 \qquad (4.168)$$

These $n + 1$ equations provide $n + 1$ values for the elements w_j of $\mathbf{w}$. The method with negative feedback as well as the least-squares method can be used only when the patterns are linearly separable, in contrast to the nearest neighbor method.

In two recent papers Pijpers applied pattern recognition techniques in order to obtain a qualitative classification of dithiocarbamate compounds from ^{13}C-NMR and IR spectroscopic data (GA 79, PJ 79). Unsupervised pattern recognition permitted qualitative classification of these complexes into five classes, denoted as free ligands, complexes of main group elements, and complexes of transition metals with both normal oxidation–coordination ratios and exceptional oxidation–coordination ratios.

Supervised techniques based on these five classes did show that the descriptors $\delta(M^{13}CS_2)$, $\nu(C \dot{=} N)$, and a parameterization of the period of the periodic system of elements, in which the complexed metal was found, were dominant in classifying compounds. Other descriptors applied were found to be less important. The classification and ranking of the descriptor quality gave an insight into the relationship between the quantities measured. Classification of other compounds with the learning machine, which disagreed with chemical expectations, were readily identified and this improved understanding of the chemical background.

Another application of pattern recognition in analytical chemistry is presented by Simon et al. (SI 77). Here a judicial problem encountered in the felonious use of paper money was solved by analyzing 19 paper samples representing 11 different types of paper from seven manufacturers. The results of the paper analysis for Cu, Mn, Sb, Cd, Cr, Co, Ag, Pb, Mg, and Fe were compared by pattern recognition techniques. It was found that only six features were needed to identify 16 of 19 papers completely. The three paper samples not completely separated came from different manufacturing runs of the same type of paper.

4.8   OPTIMIZATION

Optimization of a (chemical) system infers an adaptation of all control variables in such a manner that the best possible result is obtained. This adaptation should be within the constraints set for the control variables of the system (LO 69). A mathematical formulation of this situation could be to optimize a function φ that depends on n adaptable parameters X_i and m nonadaptable parameters Y_j:

$$\varphi = F(X_i, Y_i) \qquad (1 \leqslant i \leqslant n, \quad i \leqslant j \leqslant m) \qquad (4.169)$$

Mostly the effect of the nonadaptable parameters on φ is considered to be random, which means that the results of variations of Y_j on φ are considered to follow a Gaussian distribution function around an average value of zero. Moreover the effect of Y_i and Y_j is assumed to be independent for all i and j $(i \neq j)$. These simplifying assumptions lead to

$$\varphi = F(X_i) \qquad (1 \leqslant i \leqslant n) \qquad (4.170)$$

The graphic representation of the response of φ or variations of all x_i parameters is called a response plane. Connections between points with equal response on variation of various x_i-parameters are called equiresponse lines on the response plane. Five possible equiresponse lines for a two-dimensional response plane are depicted (DA 71) in Figure 4.30).

Sometimes the relationship between φ and x_i is known as a mathematical equation with a set of boundary conditions. In these situations mathematical techniques for searching for optimal combinations of x_i values may be used, for example in linear programming. When a mathematical relationship is unknown, experimental optimization techniques may be of help. Here a distinction can be made between simultaneous techniques and sequential ones.

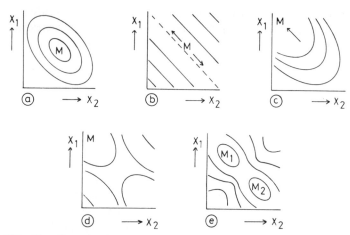

Figure 4.30. Two-dimensional representation of the shape of equiresponse lines for various situation. (*a*) One discrete extremum; (*b*) a stationary rig; (*c*) a sloping rig; (*d*) a minimax saddle; (*e*) multiple maxima (or minima).

Simultaneous Optimization Techniques

The initiation of this technique requires starting values for the x_i-parameters under investigation and a set of combinations. The various combinations of parameter values produce a response plane. With two parameters a two-dimensional (flat) plane is formed; more parameters produce a more dimensional response space that cannot be represented in a flat plane without losing information. Some schematic examples of simultaneous optimization techniques are given.

1. *Random Design.* Suppose φ to be a function of two adaptable parameters x and y. A number of experiments can be designed for a set of combinations of x- and y-values selected at random. The experiment producing the best φ-value is called the optimal condition. Another procedure based on this method starts from a set of increments for x- and y-values. The values $x_1, x_2, \ldots, x_a$ and $y_1, y_2, \ldots, y_b$ are combined in all possible pairs (x_i, y_j) with $1 \leqslant i \leqslant a$ and $1 \leqslant j \leqslant b$. From the set of all possible pairs a random selection of some pairs is made and these are used as experimental conditions. This procedure is called *simple random design*. The location of the optimum can be found with enhanced accuracy by repeating the complete procedure in the direct neighborhood of the optimum found with a first trial. The operation of the random design procedure may be improved by application of some strategy for the selection of the parameter values that fix the experimental conditions. In the situation

where two parameters should be selected the response plane is subdivided into *n* equally shaped surface parts. Within each of these parts one randomly selected set of parameter values is used for the experiment. This method is called *strategic random design optimization*; it is also applicable for optimization problems with more than two parameters. In practice this method is most useful when a large number of parameters should be selected for a procedure that yields a small experimental error.

2. *Factorial Design.* The factorial design method for optimization starts by selecting an increment that is equal for all adaptable parameters $x_1, x_2, \ldots, x_a$ and $y_1, y_2, \ldots, y_b$. For example, x_1, x_2, x_3, and x_4 are combined with y_1, y_2, and y_3. Contrary to the random design method all possible combinations (x, y) are investigated in order to find the optimal condition. Here too the accuracy is enhanced by a second trial around the optimal value with diminished incremental values. By duplicating the experiments, application of an analysis of variance scheme on the results is possible. This results in an estimate of the experimental errors involved and an estimate of the significance of the parameters under investigation. A disadvantage of this method is the large increase in the number of experiments for an increase in the number of parameters. This disadvantage may be partially circumvented by application of special strategies for parameter selection (half-factorial design).

Sequential Optimization Techniques

An interesting possibility for experimental design is offered by a procedure that does not fix the selection procedure in advance. Here a start is made from a very limited set of experimental conditions. Based on the results of these experiments one (or some) more experimental condition is selected. Of course, the methods mentioned may be combined and in practice various possibilities are applied. For instance, simultaneous method may give a global location of the optimum. Thereafter a sequential method may be applied for a more accurate location of the optimum. However, the inverse order may also used: in the neighborhood of an optimum found by a sequential method, a simultaneous optimization technique may be applied in order to verify the results and avoid false optima.

1. *Single-Step Procedures.* Here alternatively one factor is varied and all others are kept constant. The factor with the highest response for example, *y*, is investigated first, followed by the next highest, for example, *x*, and so on. A possible scheme for this method (Figure 4.31) is as follows:

    a. Estimate the coordinates of the optimal condition; in the two-dimensional situation this could be (x_0, y_0).

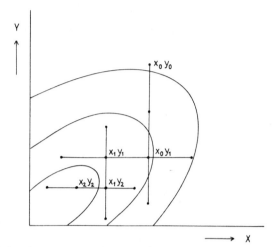

Figure 4.31. Order of measurements for a two-dimensional problem attacked with a single-step procedure.

b.   From this origin x_0 is kept constant and four equidistant y-values are selected covering the entire range of y-values. An optimum is found for, say, y_1. Sometimes a better estimation is possible by fitting a polynomial function of order 3 through the four responses, which yields a better value for y_1.

c.   A second set of experiments is carried out at y_1 and four equidistant x-values covering the entire x-range producing an optimal x_1 value.

d.   A second cycle of optimization is started for both parameters, now searching with smaller increments.

e.   The process is disrupted when for instance the change in the response is smaller than a predetermined limiting value.

2.   *Steepest-Ascent Method.*   Brooks (BR 59) gives two types of application of this method.

a.   A "slope" procedure, starting from (x_0, y_0) according to a factorial scheme, say, $(2*2)$: an estimation of the gradient based on the response of that scheme gives the direction of the steepest slope. Following this direction a new factorial scheme with interval S is constructed, producing a new estimation of the gradient. The procedure is repeated until the changes of the obtained response decrease. Finally a set of experiments is performed about the location with the optimal response, followed by a fitting procedure or a straightforward selection of the best response (Figure 4.32).

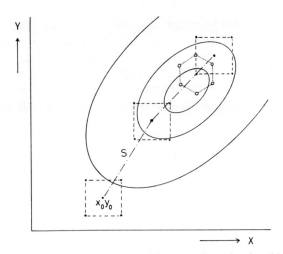

Figure 4.32. Steepest-ascent method for a two-dimensional problem.

b. Single-step method: just as with the slope method, one starts with a factorial scheme around an estimated optimum (x_0, y_0) in order to find the direction of the highest gradient g_0. The second experiment starts at a distance S from (x_0, y_0) along g_0. The magnitude of the average response at this location is compared with that at (x_0, y_0): a higher response is followed by another experiment at a distance $2S$ from (x_0, y_0) along g_0, which is repeated until the average response stops growing (or until a boundary limit of one of the parameters is met). Hereafter a second direction of the gradient g_1 is calculated, followed by new experiments along this direction. The procedure is repeated until a disrupting criterion is met. The unique estimate of the optimal factor combination stems from the best response.

3. *Simplex-Optimizations.* A simplex is a geometric construction fixed by a number of points equal to the number of dimensions of the problem space plus 1. The number of dimensions of the problem space equals the number of adaptable parameters, which in principle is not limited. The simplex method aims for a displacement of the simplex towards a region in the problem space with the optimal response. An equilateral triangle is applied for simplex optimization of a problem with two parameters. The angular points of the triangle are called vertexes (DE 73). A tetrahedral-shaped simplex is applied for three parameters, and so forth.

As is seen with the other optimization methods, usually only the most significant factors are selected for optimization. The relative importance of various factors follows from a factorial scheme. The importance of the factors may vary with the position in the problem space: around the starting location the order of importance can be quite different from that at the location of the optimum; therefore at the end of a simplex optimization sometimes a fractional scheme is constructed in order to rank the influence of all factors on the response in the neighborhood of the optimum.

The incremental distance for the factor variation is usually selected in such a way that the change in the response upon a stepwise change of a factor is about equal for all factors. At the initial stage a great increment is nearly always advantageous because of a quicker approach to the optimum. Moreover, under these conditions experimental errors play a less important role.

The boundary conditions for the factor space should be set in advance. These boundary conditions may result from limitation of temperature, pressure, solubility, and instability.

In order to position the first simplex the starting values of all factors should be selected. In practice these values often result from previous experiments. The starting conditions do not affect the efficiency of the method in a significant way; however, if extra factors should be included at a later stage it is advantageous to position one of the sides of the equilateral triangle parallel to the axis of one of the factors (see LO 69, SP 62, YA 74). Of course this is useful only when a visual inspection of a two-dimensional problem is made. Problems of higher dimensionality should preferably be handled with a computer. A detailed description of the order of the steps to be taken, simplex calculations, and warnings about pitfalls that may occur is found in a review article of Deming and Parker (DE 78).

A comparison of the simplex method with respect to other methods can be summarized as follows:

Advantages
- A relatively small number of experiments to be made.
- Simple calculations.

Disadvantages
- No complete certainty about the absolute optimum. This also holds for other techniques; however, screening off by means of a suboptimum does not work for simultaneous techniques.
- Handling a dimensionality greater than two rules out visual inspection and makes recognition of cycling around an optimal value hard to detect.

TABLE 4.20. Comparison Between Simultaneous and Sequential Optimization Methods

Simultaneous Methods	Sequential Methods
A scheme for a great number of experiments is fixed in advance	A limited set of experiments is fixed in advance. Based on the responses of this first set of experiments the next set is designed
A response criterion (the density of the experiments in the experimentation space) is set in advance	—
After all experiments of the scheme are finished, the optimum is selected from the responses. Sometimes a second scheme is designed in the neighborhood of this optimum	The series of experiments is disrupted when a criterion, set in advance, is met (this can be a minimal change in response during a fixed number of experiments)
Almost the entire response plane is scanned	Only a part of the response plane is scanned, based on an estimated optimum and moving in the direction of the most advantageous response
Erroneous high results of experiments easily result in false optima. To avoid this, experiments should be duplicated or a second scheme should be investigated	Erroneous high results are detected because of their prolonged effect on a great number of consecutive experiments and the resulting optimization procedure. The result is corrected by duplication of the experiment.
Low accuracy of the optimum found	Mostly the accuracy of the optimum is better
Well applicable to experiments with prolonged duration	Not well applicable to experiments with prolonged duration
Hardly any influence at all on the optimization of experimental errors	The results are error prone
False optima may be found when (too) high incremental values are applied	False optima may be found with (too) big steps and with screening off
	The technique may be automated. A graphic method can be performed without calculations (in the two-dimensional situations)

• The original method has no provisions for hastening the process, however, see MO 74 for a modified procedure that copes with this.

A summary of all methods discussed is given in Table 4.20. Brooks (BR 59) compares various techniques, augmented by Spendley (SP 62), with the simplex method. It should be noted here that in the experiments given by Brooks, some refinements on the steepest-ascent procedures have been applied, in order to avoid difficulties due to experimental errors. The simplex procedure suffers from lack of such refinements. A more realistic basis for comparison is the selection of the 95% of all experiments that end normally. The comparison based on this selection is given in parentheses in Table 4.21.

As an example one may consider the chromotropic acid method for the determination of formaldehyde, after optimization with the simplex method, of Houle (HO 70). Starting from a method described by other investigators, he obtained a 36-fold increase in sensitivity over the original procedure. The optimized method is also much faster (6 minutes versus 45 minutes) and is at least as precise.

Czeck (CZ 73) described the optimization of the J-acid method and the acetylacetone method for the determination of formaldehyde, starting with published methods. The J-acid method turned out to be five times as sensitive as the original procedure after the analytical parameters were selected by a simplex procedure. After optimization the acetylacetone method was almost four times as sensitive as the original AOAC method and a tenfold increase in analytical speed was obtained. The papers by Czeck describe the procedure and its constraints in detail.

TABLE 4.21. **Comparison of Various Optimization Techniques Based on Mean Values of Responses (SP 62)[a]**

Method	Without Experimental Errors		With Experimental Errors	
	$N = 16$	$N = 30$	$N = 16$	$N = 30$
Factorial design	0.9542	0.9671	0.9541	0.9602
Single step	0.9696	0.9842	0.9588	0.9726
Steepest ascent				
Single	0.9652	0.9866	0.9489	0.9718
Slope	0.9902	0.9960	0.9755	0.9846
Random design				
Single	0.8706	0.9642	0.8851	0.9415
Strategic	0.9345	0.9590	0.9243	0.9541
Simplex	0.9813	0.9934	0.9442	0.9191
			(0.9711)	(0.9716)

[a]The maximal attainable response is 1.0; N = number of experiments.

4.9 COMPUTATIONAL ASPECTS

The application of computer methods to computations with analytical data starts with feeding the data into a computer-accessible memory device. In situations where only few data should be fed into the computer, punch cards can be used. The advantage of punch cards is that the data may be kept "forever" at the disposal of the analyst and can be repeatedly used. A disadvantage may be that computer cards are produced by humans and thus are error prone. This disadvantage was realized long ago and a system of blind typing was developed to eliminate typing errors by punching with two independent operators. One operator punches the actual cards and a second one punches on a device that signals whenever disagreement occurs between the first and second typing.

In situations where large amounts of data are produced by an analytical instrument, a device that produces paper tapes may be useful. Here the analytical data are converted into a code of binary digits and a set of data is converted within a relatively short period into a paper tape. The advantage of a paper tape being capable of carrying large amounts of data on a cheap medium is sometimes canceled by the fragile structure of the data carrier. Depending on the amount of data produced and the speed of the signal flow, magnetic devices such as tapes and disks can be of use. A magnetic tape is capable of carrying large data sets (800–1600 bits/in.) on six or nine tracks and can be fed with much higher speeds than paper tapes. Here a disadvantage may be that the data should be read and written in order, and to read two data, one at the beginning and one at the end of the tape, requires complete unwinding, which is time-consuming. In situations where data are produced at regular time intervals and should be processed between two production intervals, this is prohibitive for tape applications. Here random-access devices such as disks should be used. A pulsed NMR spectrometer capable of Fourier transforming the measured signal decays operates with a disk. The disadvantage of a disk is that it has a limited capacity in comparison with a tape; therefore it should preferably not be used as a storage device for information.

Other situations arise when, apart from various analytical instruments, large computer facilities are available. Here networks may be constructed that feed the computer memory banks at regular time intervals with information produced by analytical instruments. Two different situations may exist.

• The computer, as a master, asks each of the instruments at regular intervals for information. The instruments are "slaves" that produce information at the request of the computer.

• The instrument signals the computer that a set of measurements has been completed and will be transferred to the computer memory. Here the instrument sets the priority and acts as a master, using the computer as a slave.

When microprocessors and memory become cheaply available analytical instruments may be extended with their own private memories for temporary storage of a limited set of information until the master computer is ready to consume it. A general discussion of all possible situations is beyond the scope of this book, and special equipment such as optical mark readers and optical curve followers are not discussed here.

REFERENCES

AG 71 R. B. d'Agostino: *Biometrika* **58**, 341 (1971).

AM 77 A. E. Ames, G. Szonyi: in *ACS Symposium Series No. 52* American Chemical Society, Washington, D.C., (B. R. Kowalski, Ed.), 1977, pp. 225–227.

AN 72 Harry C. Andrews: *Introduction to Mathematical Techniques in Pattern Recognition*, Wiley-Interscience, New York, 1972.

BO 64 G. E. P. Box, D. R. Cox: *J. R. Stat. Soc.*, **B26**, 211 (1964).

BO 70 G. E. P. Box, G. M. Jenkins: *Time Series Analysis (Forecasting and Control)*, Holden-Day, San Francisco, 1970.

BR 59 S. H. Brooks: *Oper. Res.* **7**, 430 (1959).

BR 62 R. G. Brown: *Smoothing, Forecasting and Prediction of Discrete Time Series*, Prentice-Hall, New York, 1962.

CL 79 P. Cleij, A. Dijkstra: *Z. Anal. Chem.* **298**, 97 (1979).

CO 71 W. J. Conover: *Practical Nonparametric Statistics*, Wiley, New York, 1971, p. 302.

CZ 73 F. P. Czeck: *J. AOAC*, **56**, 1489, 1496 (1973).

DA 71 O. L. Davies: *Design and Analysis of Industrial Experiments*, Oliver and Boyd, Edinburgh, 1971.

DA 72 O. L. Davies, P. L. Goldsmith: *Statistical Methods in Research and Production*, Oliver and Boyd, Edinburgh, 1972.

DE 73 S. N. Deming, S. L. Morgan: *Anal. Chem.* **45**(3), 278A (1973).

DE 75 D. L. Duewer, B. R. Kowalski, T. F. Schatzki: *Anal. Chem.* **47**, 1573 (1975).

DE 78 S. N. Deming, L. R. Parker: "C.R.C.-Critical Reviews in Analytical Chemistry," **7**, 187 (1978).

DI 69 W. J. Dixon, F. J. Massey: *Introduction to Statistical Analysis*, 3rd ed., McGraw-Hill, New York, 1969, p. 243.

DI 80 C. B. M. Didden, H. N. J. Poulisse, "On the Determination of the Number of Components from Simulated Spectra using Kalman Filtering," *Anal. Lett.* **13** (A11), (1980).

DP 75 F. Dupuis, A. Dijkstra: *Anal. Chem.* **47**, 397 (1975).

DU 73 R. O. Duda, P. E. Hart: *Pattern Classification for Science Analysis*, Wiley-Interscience, New York, 1973.

DU 75 D. L. Duewer, B. R. Kowalski: *Anal. Chem.* **47**, 526 (1975).

DW 75 D. L. Duewer, J. R. Koskinen, B. R. Kowalski: *Arthur Pattern Recognition Manual*, University of Washington, Laboratory for Chemometrics, Department of Chemistry, 1975.

DY 74 A. Dyer: *Biometrika* **61**, 185 (1974).

EC 71 K. Eckschlager: *Coll. Czech. Chem. Commun.* **36**, 3016 (1971).

EC 72 K. Eckschlager: *Coll. Czech. Chem. Commun.* **37**, 137 (1972).

EC 73 K. Eckschlager: *Coll. Czech. Chem. Commun.* **38**, 1330 (1973).

EK 72 K. Eckschlager: *Coll. Czech. Chem. Commun.* **37**, 1486 (1972).

FR 66 R. D. B. Fraser, E. Suzuki: *Anal. Chem.* **38**, 1770 (1966).

FR 69 R. D. B. Fraser, E. Suzuki: *Anal. Chem.* **41**, 37 (1969).

GA 79 H. L. M. van Gaal, J. W. Diesveld, F. W. Pijpers, J. G. M. van der Linden, *Inorg. Chem.* **18**(11), 3251 (1979).

GO 76 Z. Govindarajulu, *Comm. Stat. Theory Methods* **A5**, 429 (1976).

GO 72 G. Gottschalk: *Z. Anal. Chem.* **258**, 1 (1972).

GR 77 L. A. Gribov, M. E. Elyashberg, V. V. Serov: *Anal. Chim. Acta* **95**, 75 (1977).

HA 74 A. den Harder, L. de Galan: *Anal. Chem.* **46**, 1464 (1974).

HE 72 G. M. Hieftje: *Anal. Chem.* **44**(6), 81A (1972).

HI 72 G. M. Hieftje: *Anal. Chem.* **44**(7), 69A (1972).

HO 70 M. J. Houle, D. E. Long, D. Smette: *Anal. Lett.* **3**, 401 (1970).

HO 71 G. Horlick: *Anal. Chem.* **43**(8), 61A (1971).

HY 70 Hyunyong Kim: *J. Chem. Ed.* **47**(2), 120 (1970).

JU 75 P. C. Jurs, T. L. Isenhour: *Chemical Applications of Pattern Recognition*, Wiley, New York, 1975.

KA 70 H. Kaiser: *Anal. Chem.* **42**(2), 24A (1970).

KA 61 R. E. Kalman, R. S. Bucy: *J. Basic Eng. (ASME)* **83D**, 95 (1961).

KA 74 B. R. Kowalski, C. F. Bender: *J. Am. Chem. Soc.* **96**, 916 (1974).

KI 70 H. Kaiser: *Anal. Chem.* **42**(3), 26A (1970).

KL 75 L. A. Klimko, C. E. Antle: *Commun. Stat.* **4**, 1009 (1975).

KO 71 B. R. Kowalski, C. A. Reilly: *J. Phys. Chem.* **75**, 1402 (1971).

KO 72 B. R. Kowalski, T. F. Schatzki, F. H. Stross: *Anal. Chem.* **44**, 2176 (1972).

KO 73 B. R. Kowalski, C. F. Bender, *J. Am. Chem. Soc.* **95**, 686 (1973).

KO 74 B. R. Kowalski: *Pattern Recognition in Chemical Research in Computers in Chemical and Biochemical Research*, Vol. 2, Academic Press, New York, 1974.

KW 72 B. R. Kowalski, C. F. Bender: *J. Am. Chem. Soc.* **94**, 5632 (1972).

KW 74 B. R. Kowalski: *Chem. Tech.* **1974**, 300 (May).

LI 67 H. W. Lilliefors: *J. Am. Stat. Assoc.* **62**, 399 (1967).

LI 78 C. B. G. Limonard, F. W. Pijpers: *Anal. Chim. Acta.* **103**, 253 (1978).

LO 69 D. E. Long: *Anal. Chim. Acta* **46**, 193 (1969).

MA 51 F. J. Massey, Jr.: *J. Am. Stat. Assoc.* **46**, 69 (1951).

MA 73 D. L. Massart: *J. Chromatogr.* **79**, 157 (1973).

MO 74 S. L. Morgan, S. N. Deming: *Anal. Chem.* **46**, 1170 (1974).

MS 78 P. J. W. M. Müskens: Thesis, Catholic University of Nijmegen, 1979, p. 84, IV 3.

MU 78 P. J. W. M. Müskens: *Anal. Chim. Acta* **103**, 445 (1978).

OW 62 D. B. Owen: *Handbook of Statistical Tables* Addison-Wesley, Reading, Mass., 1962, pp. 413–425.

PE 67 J. Peters: *Einführung in die Informations Theorie*, Springer, Berlin, 1967.

PI 67 J. Pitha, R. M. Jones: *Can. J. Chem.* **45**, 2374 (1967).

PJ 79 F. W. Pijpers, H. L. M. van Gaal, J. G. M. van der Linden: *Anal. Chim. Acta* **112**, 199 (1979).

PO 79 H. N. J. Poulisse, *Anal. Chim. Acta* **112**, 361 (1979).

RE 74 G. Redl, R. D. Cramer, C. E. Berkhoff: *Chem. Soc. Rev.* **3**, 273 (1974).

SC 73 E. I. Schuster: *J. Am. Stat. Assoc.* **68**, 713 (1973).

SE 76 P. F. Seelig, H. N. Blount: *Anal. Chem.* **48**, 262 (1976).

SE 79 P. F. Seelig, H. N. Blount: *Anal. Chem.* **51**, 327 (1979).

SH 49 C. E. Shannon, W. Weaver: *The Mathematical Theory of Communication*, University Press of Illinois, Urbana, 1949.

SH 65 S. S. Shapiro, M. B. Wilk: *Biometrika* **52**, 591 (1965).

SH 68 S. S. Shapiro, M. B. Wilk, J. H. Chen: *J. Am. Stat. Assoc.* **63**, 1343 (1968).

SH 72 S. S. Shapiro, R. S. Francis: *J. Am. Stat. Assoc.* **67**, 215 (1972).

SI 56 S. Siegel: *Nonparametric Statistics for the Behavioral Sciences*, McGraw-Hill, New York, 1956, p. 44.

SI 77 P. J. Simon, B. C. Giessen, T. R. Copeland: *Anal. Chem.* **49**, 2285 (1977).

SL 73 J. J. Schlesselman: *J. Am. Stat. Assoc.* **68**, 369 (1973).

SM 70 H. C. Smit: *Chromatographia* **3**, 515 (1970).

SM 80 H. C. Smit: *Anal. Chim. Acta* **122**, 201 (1980).

SO 65 D. P. Shoemaker, C. V. Garland: *Experiments in Physical Chemistry*, 2nd ed., McGraw-Hill, New York, 1965.

SP 62 W. Spendley, G. R. Hesel, F. R. Hinsworth: *Technometrics* **4**, 441 (1962).

ST 68   M. A. Stephens, M. R. Maag: *Biometrika* **5**, 428 (1968).

ST 70   M. A. Stephens: *J. R. Stat. Soc. Ser. B*. **32**, 115 (1970).

ST 74   M. A. Stephens: *J. Am. Stat. Assoc.* **69**, 730 (1974).

ST 62   H. Stone: *J. Opt. Soc. Am.* **52**, 998 (1962).

VA 75   B. G. M. Vandeginste, L. de Galan: *Anal. Chem.* **47**, 2124 (1975).

WE 69   A. W. Westerberg: *Anal. Chem.* **41**, 1770 (1969).

YA 74   L. A. Yarbro, S. N. Deming: *Anal. Chim. Acta* **73**, 391 (1974).

ORGANIZATION

5.1 PLANNING

One of the quality aspects of analytical chemistry is the optimal execution of analytical tests with respect to time and money. As is shown in the preceding chapters much can be done by using the appropriate techniques for estimating the optimal sampling frequency, by selecting the optimal analytical method, and by applying optimal data handling. However, there are additional ways to influence the time and cost of analytical procedures. Much time and money can be wasted by improper planning of such aspects as availability of instruments, reagents, and people and improper routing, causing congestion and improper sequencing of actions. Much time and money can be gained by having reliable forecasts of availability of new methods and instruments and estimations of the probability that new methods will be found.

All these factors can be included under the heading "planning." Planning in this context means arranging possible actions with respect to quality, quantity, and sequence. The basic objective of planning is to minimize time or cost and to maximize resource utilization. Planning thus sets the boundary conditions of a network of interrelated parts of the work to be done, the quantification of the size of the network links in time or money, and the reliability of the quantification, and denotes ways for selecting the optimal path.

Planning can be subdivided into forecasting, evaluation, and control. In this respect it resembles a common control problem and in fact it should be considered as such. In practice, however, many of the available methods are static and are used in a static way. A host of static and dynamic planning procedures can be found in the area of operations research (OR). In this chapter we discuss the concepts of OR from the point of view of analytical chemists.

5.1.1 Forecasting Methods

When one decides what work will or must be done in the future, there is a need for forecasting. Here one may be thinking in terms of some different time scales, the short, the medium, and the long term. In the short term, at least, one is heavily constrained by an existing work load, but even here there can be an element of choice. Typically the work input exceeds one's capacity; thus a selection of tasks has to be made and, within that selection, an order of priority assigned. As the time scale increases, one's freedom of choice increases, at least in analytical work. One's choice is generally maximal in the longer term, and it is here that some of the forecasting techniques may be applied. There is a variety of approaches, but only a few that are applicable to analytical work and analytical laboratories are mentioned here.

Trend Extrapolation

One assumes that the advances in a particular field will follow a defined pattern or progress in a relatively orderly manner, the so-called surprise free projection. As a rule this extrapolation, for example, analysis time, analysis cost, and limit of detection, is performed in a quantitative way. With the aid of curve fitting (Section 4.4.1) the descriptor under consideration is described by an equation that fits the available data best. When the equation has been obtained, it is quite easy, although very dangerous, to extrapolate into the future.

The development of many analytical techniques during the last decades has been such that an exponential curve seems to fit the data, for example, for the number of publications on analytical techniques (Figure 5.1) (BR 75, BR 76, BR 80). Maybe this is true, but not for the extrapolation. A more realistic curve for technical development is an S-shaped curve that levels off, resembling the first part of the well-known growth curve of microorganisms. However, the prediction curve of a particular technique can be connected to another curve of another newly developed technique that surpasses the former one. An illustration is the series of fitted curves on the number of papers devoted to separation methods (Figure 5.2).

A rough estimation of the speed or precision of subsequent colorimetric and spectrophotometric techniques will probably show the same trend. Separation methods such as analytical distillation, gas chromatography, and high-performance liquid chromatography are other examples. For short-term extrapolation the method can be very useful, allowing estimations of, say, future work load, and thereby allowing predictions of the time when the use of automatic analyzers will be feasible. Here decreasing cost per analysis with an increase in work load allows estimations of cost in the future, when combined with trend analysis of labor cost.

Another trend extrapolation method is time series analysis combined with predictive filtering methods, for example, Kalman filtering (Section

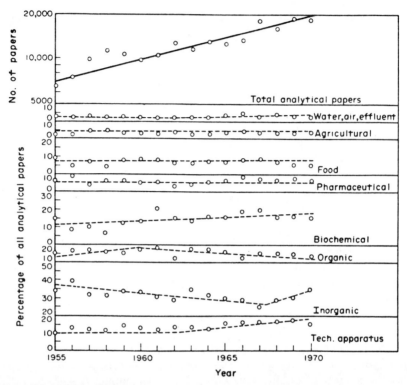

Figure 5.1. Development of analytical chemistry showing broad trends for eight categories of analytical chemistry for the period 1955–1970. Data expressed as percentage of total number of analytical papers (BR 75).

4.3.2). This is a typical short-term method that has its merits in describing and forecasting analytical quality parameters, in particular when these are influenced by more or less random factors, for example, complaints from customers.

Needs Analysis

With this method one examines the needs expected in the future and estimates the probability that the necessary technology or expertise will be available in time. One of the applicable methods is the writing of a so-called surprise free scenario. Here one describes all possibilities in a logical and justified way, estimating the probability of the possible alternatives. The "relevance tree" is pursued until all possibilities above a certain probability level for a predetermined time span have been described.

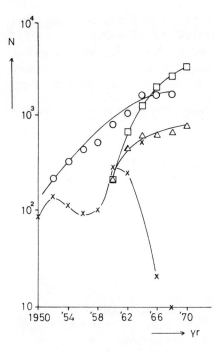

Figure 5.2. Development of analytical methods (BR 75). (x) Analytical distillation; (o) cumulative analytical distillation; (Δ) gas chromatography; (□) cumulative gas chromatography.

This method, which has been described in detail by Kahn and Wiener (KA 67), has been developed for such important forecasts as the economic development of countries. It is also used in a rudimentary form by many managers to explain why they need more instruments or equipment in the future. It seems worthwhile to develop this method for analytical purposes and to publish the results in the analytical literature.

Dynamic Modeling

Dynamic modeling comprises the use of mathematical models, analogue or digital, to simulate situations and interactions expected in the future. These models are now being employed extensively in economic forecasting. One of these models has gained much attention as the "world model" (FO 71), revealing the future situation in world economics, population, and pollution, for example.

Until now only a few examples of dynamic models in analytical chemistry have been described. Schmidt (SC 75, SC 77), Vaananen et al. (VA 74), and Doerffel (DO 75) applied digital simulation and queueing theory to analytical laboratory systems. Van de Akker and Kateman (AK 76) described a simulation model of an analytical laboratory to be used as a simulation game in the education of analytical chemists. Vandeginste described a hardware laboratory model (VA 77).

The most complete account of this field has been given by Vandeginste (MA 78, VA 79, VA 80), who applied queueing theory to a laboratory organization and constructed a model that simulates—and resembles—the operation of a real spectroscopic laboratory for structural analysis of organic compounds. Because the delay of samples in an analytical laboratory usually exceeds the analysis time, most of the time is spent waiting. Here queueing theory may help to enhance some of the quality parameters of the laboratory: analysis delay time and cost. Vandeginste (VA 79), restricting himself to a spectroscopic laboratory, stated that a dynamic planning strategy should answer the following questions:

- Which strategy (or decision rules) should be followed to select the analytical method for solving the structure of the compound, taking into account the estimated probability that the analytical methods might solve the problem and minimize the queue lengths in the laboratory?
- What should be the procedure when an analytical method fails to solve the structure of the compound under investigation? After what analysis time should the analysis be terminated and another method tried?
- Which priority rules should be applied and how do they affect the waiting time? (For example, priority between samples of various origin, between samples that were unsuccessfully analyzed, and between easy and difficult samples.)
- Which kind of priority rule (absolute or relative) should be applied?
- What is the effect of interruptions of the analytical procedure by other activities on the waiting time?
- What is the influence of the structure of the organization, for example, centralized versus decentralized?
- Is it advantageous to accept samples only at discrete intervals?

Queueing theory answers the above questions for simplified models only. However, from theoretical calculations on simple systems, the effect of the variables mentioned may be estimated for more complex systems. Vandeginste mentions the following results. The utilization factor (ρ) of the service channel plays an important role in all kinds of service systems, and an analytical laboratory is such a system. It is defined as

$$\rho = \frac{\overline{AT}}{\overline{IAT} \cdot m} \tag{5.1}$$

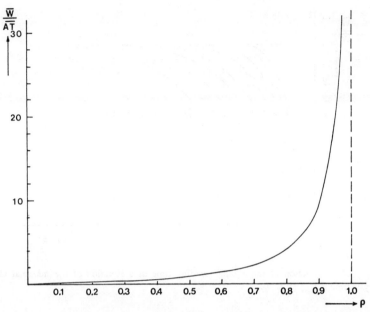

Figure 5.3. The ratio between the average waiting time ($\overline{W}$) and the average analysis time ($\overline{AT}$) as a function of the utilization factor (ρ) for a system with exponentially distributed interarrival times and analysis time (M/M/1 system) (VA 79).

with $\overline{AT}$ = mean analysis time
$\overline{IAT}$ = mean interarrival time
m = number of analysts serving the channel
For simple systems (the so-called M/M/1 systems*) where the samples are analyzed in the sequence of arrival at the system [first in, first out (FIFO) rule] the mean waiting time $\overline{W}$ is given by

$$\overline{W} = \frac{\overline{AT}^2}{2(1-\rho)\overline{IAT}} \tag{5.2}$$

with $\overline{AT^2}$ = second moment of the analysis time.

The asymptotic shape of Figure 5.3 is characteristic for all queueing systems. From this figure it is seen that for values of $\rho > 0.85$, minor variations in the organization of activities in the laboratory may induce a severe change in the waiting time.

*In an A/B/C system, A depicts the type of arrival process, B depicts the type of service times, and C depicts the number of service channels (analysts). M means Markovian: each event is independent of the others.

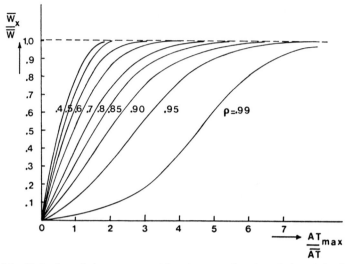

Figure 5.4. Reduction of the average waiting time as a function of the maximal allowed analysis time AT_{max} (VA 79).

It can be shown that decreasing the coefficient of variation of the analysis time—by processing many samples of the same kind for instance—decreases the waiting time, although the effect is less than the effect of a decrease of the mean analysis time, for example, by speeding up analyses. Figure 5.4 demonstrates that truncating the analysis and establishing a certain ratio between maximum analysis time and mean analysis time can improve the waiting time considerably, if other channels are free. Otherwise the reverse effect may manifest itself. The effect of the mean interruption time β and the mean time between interruptions α (e.g., other activities while samples are waiting or breakdown of instruments) is given by

$$R = \frac{1-\rho}{\alpha/(\alpha+\beta)-\rho} \tag{5.3}$$

with $R=$ ratio of mean waiting times with and without interruptions α and β constants. This effect is depicted in Figure 5.5.

The priority given to samples can be effected in several ways. The effect of the various rules can be approximated. For priority groups with different mean waiting times the overall lowest mean waiting time is found when

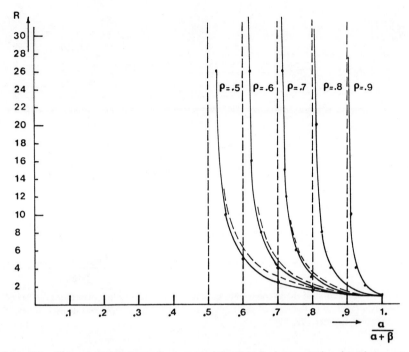

Figure 5.5. The ratio (R) between the average delay time for interrupted and not interrupted analyses, as a function of the available time for analysis. (——) α and β exponentially distributed; (-----) α and β constants (VA 79).

samples with the shortest analysis time are given absolute priority. Figure 5.6 shows the effect of this discipline as a function of the ratio of analysis times and number of samples in both classes.

In the analytical laboratory, this situation is met when an analyst executes two different analyses, or when the samples can be subdivided into two groups, for example, "easy" and "difficult" samples with "small" and "large" analysis times, respectively. Often, however, it is not possible to predict whether, say, a spectrum will be difficult to interpret or not. Thus after a certain interpretation time, the difficult spectrum may be transferred to a pile of unfinished spectra and the measurement of the next sample or interpretation of the next spectrum started. This problem can be solved only by simulation. Vandeginste proved that in spectroscopic analysis, waiting time can be improved considerably only by giving priority to the "easy" samples that have been recognized as such beforehand.

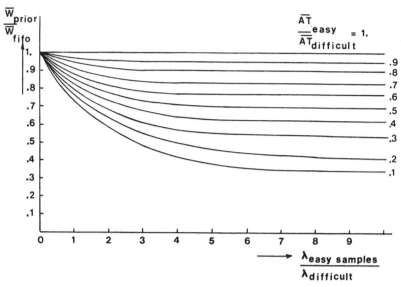

Figure 5.6. Reduction of the average waiting time, attributing absolute priority to the "easy" samples as a function of the ratio of the analysis time and ratio of the input density of "easy" and "difficult" samples (VA 79).

The results of the application of queueing on analytical laboratories can help in establishing rules for optimizing waiting time and cost in laboratories. However, in complex systems such as analytical laboratories, consisting of multiserver nodes and governed by state-dependent decision rules, the effect of the variables mentioned can only be calculated from simulations with a model of the laboratory under consideration. Vandeginste (VA 79, VA 80) describes such a model of a spectroscopic laboratory.

Intuitive Modeling

Here one attempts to assess and refine the informed opinion of one or more experts. Although the intuitive approach is applied by most people, its more formal approach has not found widespread use in analytical chemistry. One of the methods available is the Delphi technique (DA 63). This technique depends on the individual views of experts on future developments in a particular field. The reasoning behind this method is that, apart from the better position of experts to forecast future developments, it may also be assumed that when the right experts are asked to participate, they may shape the future by developing or by checking developments.

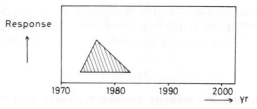

Figure 5.7. Results of a Delphi investigation in 1968 (HA 69) on the initial question, "After what year will large-scale information retrieval for science be available?" The corners of the triangle indicate first quartile, median, and third quartile.

This technique applies a carefully designed program of sequential individual interrogations, interspersed with information and opinion feedback derived by computed consensus from the earlier parts of the program. Some of the questions may, for instance, pertain to the "reasons" for previously expressed opinions. A collection of such reasons may be presented afterward to each respondent in the group, together with an invitation to reconsider and possibly revise his earlier estimates. Both the inquiry into the reasons and the subsequent feedback of the reasons adduced by others may serve to stimulate the experts to reconsider features they may have inadvertently neglected and to give due weight to factors they were inclined to dismiss as unimportant on first thought. The consensus opinions, refined by this iterative process, are often produced in graphic form. In Figure 5.7 only a small part of a much larger Delphi exercise reported by Hall is shown (HA 69). The shaded area indicates the quartiles and median of the answers.

The Delphi method operates only on the ideas the experts have in mind. Methods of widening the portfolio of ideas that one has available have been proposed. Two of the most important are lateral thinking and Synectics®. A review of methods on increasing creativity can be found in RI 74 and SO 77.

Lateral Thinking

This is the expression used by de Bono (BO 71) to describe a particular technique for enhancing individual creativity. It is shown that creative thinking consists of two consecutive stages, a concept stage followed by a processing stage. The processing stage, involving logical analysis, perception, selection, and judgment takes most of the time. It has to be restricted if one is to give the concept stage maximum scope. In order to avoid overemphasis on the processing stage, de Bono suggests that more time be devoted to restructuring and rearrangement of the original problem and to inserting discontinuities into the processing stage of one's thinking. In addition, avoiding the words "yes" and "no," for example, by application of the so-called "intermediate impossible" can be helpful.

Synectics®

Similar techniques are used by Gordon (GO 61), who coined the word synectics® for them. The essence of these methods is the application of parallel or lateral ways, by considering analogies from other disciplines for the problem under survey, paraphrasing common solutions, and using "random" ideas. One of the most widely known, but as a rule badly understood, methods of generating random ideas is that of "brainstorming." This technique requires careful preparation for its successful implementation. The members of the (semipermanent) team should be selected with care. In addition, there are specific requirements for the composition of the teams and their operations. For instance, the ideas generated should be progressed only to a preliminary stage prior to further evaluation by others.

5.1.2 Evaluation Methods

Once a reliable estimate of the future has been obtained or a set of alternative approaches of a problem is available in one way or another, it is desirable to evaluate the alternatives. An ideal way of planning, at least aesthetically pleasing, is to pursue all possibilities. In practice, however, it is more realistic to drop the least promising ways and focus on the best. In such evaluations, the application of a decision tree may be helpful. Decision trees, according to Magee (MA 64), are a means of presenting sequentially the consequences of alternative choices. Goulden (GO 74) presented an example of a decision tree shown in Figure 5.8. The example starts from the three options open to the analyst at the first decision stage. Success or failure of the options selected, together with the competitor activity (if any), will subsequently determine the options that stem from the second decision stage, and so on. One may also extend the tree with the probability of success in order to provide a measure of quantification that can be of further assistance in the decision making.

Often the system will be overprecise in an area that may have considerable uncertainty. This led Allen (AL 68) to another approach that generates a credibility bandwidth. For example, if the chance of success before a given time t_1 is rated zero, total success is considered most likely to be attained between times t_2 and t_3 and it is thought inconceivable that nothing will have been achieved at time t_4.

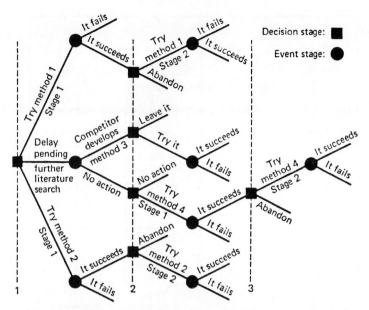

Figure 5.8. Decision tree for analytical method development (GO 74).

Massart (MA 75, MA 77) has demonstrated the use of OR techniques in analytical chemistry for evaluating alternatives in analytical procedures.

5.1.3 Control

The final stage in planning is planning control. This can be achieved in numerous ways but only a few are mentioned here.

Checklists

The most primitive, but useful, aids in control of planning are checklists. One prepares a list of actions that should be performed before the ultimate goal will be reached. During the procedure, regular consultation of the checklist prevents deviations from the desired way. These checklists can be made more or less complex by introducing order, for example, arranging actions in the sequence that is optimal or in order of importance to the final result.

Gantt Charts

A checklist in graphic form is the Gantt chart, the well-known bar representation of activities on a time scale. The graph shows a number of

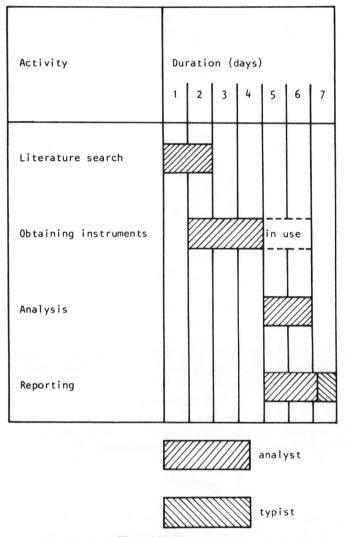

Figure 5.9. Gantt chart.

activities, the anticipated date of completion of these activities, and the means to be used (Figure 5.9).

As is shown in the example, the chart not only displays the checklist, in this example a very short one, but also gives insight into the occupation of instruments and the need of external services, for example, typists. Many layouts of such charts are available through firms that supply managers with planning aids. The host of gadgets, multicolored plastic fiches, and vast planning boards tends to obscure the usefulness of this method. Particularly in the more routine areas, where the time required for the successful completion of individual tasks can be predicted with fair certainty, the Gantt chart method can aid in the control of complex operations.

Network Diagrams

A control system more advanced than the bar chart is the network diagram. Two main types have been developed, PERT and CPM. In essence their principles are equal. The planning of a project with a network can be divided into three stages:

1. Subdividing the project into a number of separate unit operations and rearrangement into a network.
2. Estimating the time required for each unit operation and setting the critical path—the shortest possible way—through the network.
3. Pinpointing the critical points and delays and taking precautions to avoid or diminish these obstacles.

The basic rules for establishing a planning network are quite simple. The planning network can be drawn and constructed by simple arithmetic. For more complicated jobs, and the by-then very complicated network, a number of computer programs are available that perform the same tasks. Most computer programs provide the user with plotted outputs.

The procedure for establishing the network is as follows:

- Each unit operation or action is represented by an arrow. The length of the arrow in not important.
- The start and completion of an action are indicated by numbered circles, called junctions.

Each action arrow and its junctions can be drawn when the following questions are answered:

- Which action precedes the one under consideration?
- Which action proceeds simultaneously with this one?
- Which action follows this one?
- What defines the completion?

A complete network has only one starting point and one terminal.

Numbering of the junctions can be done as follows:

1.   Find the junctions without incoming arrows (with only outgoing arrows). Number these consecutively in the order in which they are found.
2.   Delete all arrows that depart from numbered junctions.
3.   Apply rules 1, 2, 3, etc.

The network can be completed by establishing the time of completion of each action. Now it is possible to indicate at each junction the earliest possible time of completion by adding to the preceding earliest possible time, the time indicated by the incoming arrows. The highest value obtained is noted at the junction. The last allowable moments are obtained by working backward in the same way as used when estimating the earliest possible time. The difference of both times is the idle time for that junction. The path along the junctions without idle times is the critical path. This determines the shortest possible time for completion of the whole task.

Figure 5.10 shows a very simple network, and Figure 5.11 shows an alternative way of drawing this network, which allows a quick survey of all idle times and the critical path.

Grubbs (GR 62) indicated the conditions to compute the expected time of completion of the whole task:

$$t_E = \frac{A + 4M + B}{6} \tag{5.4}$$

with A = most optimistic estimation of time of completion
M = most probable estimate of time of completion
B = most pessimistic estimation of time of completion
The variance of this estimate is

$$V(t_E) = \frac{(B - A)^2}{36} \tag{5.5}$$

The standard deviation is then

$$\sqrt{V(t_E)} = \frac{|B - A|}{6} \tag{5.6}$$

It must be emphasized that the assumptions used in deriving these equations are not generally accepted. However, the equations provide acceptable accuracy in practice. When a 3σ level is accepted for the probability that an estimate will occur in practice, then it is clear that the expected

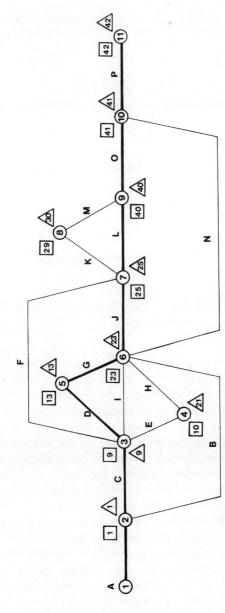

Figure 5.10. PERT network planning. *A*, sample arrives; *B*, preexamine sample; *C*, survey literature; *D*, order chemicals; *E*, select method; *F*, borrow instrument; *G*, read technical information; *H*, prepare method; *I*, wait for chemicals; *J*, test method; *K*, prepare sample; *L*, order instruments; *M*, prepare test samples; *N*, wait for instruments; *O*, test instruments; *P*, measurement; ○ most likely time (days); □ earliest possible moment; △, latest possible moment; —, critical path.

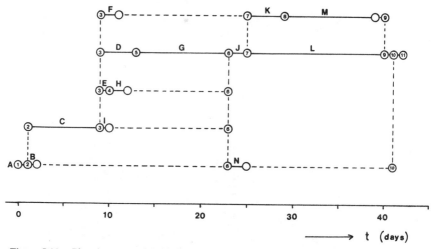

Figure 5.11. Planning network with linear time scale. —, Active time; ---, passive time. For explanation of symbols see Figure 5.10.

time of completion will have a value between A and B (as might be expected!).

Flow Charts

The extensive use and acceptance of flow charts in analysis and planning of computer programs has led to the development of the flow chart technique for planning in (analytical) research. Although Davies (DA 70) was the first one to publish a proposal, the method seems to have been developed independently in several places. In the diagram actions are represented by rectangles, decision points by diamonds, interdependent points by circles, and end points (failure or success) by rounded rectangles. The diagram has the advantage that recycling and feedback sequences can be shown. These activities are an essential part of research activity. Quantification of probabilities and times for the activities and decisions is possible and allows the estimation of probabilities of failure or success and of time of completion. For the more complicated jobs computer simulation is feasible. In Figure 5.12 a flow chart, derived from one first published by Gottschalk (GO 72), depicts an analysis of an unknown sample. When the times of completion are used as shown in Figure 5.12 and the probabilities of the decision steps are estimated, Figure 5.13 can be constructed. It is clear that the total time T is given by

$$T = t_i + nt_r \tag{5.7}$$

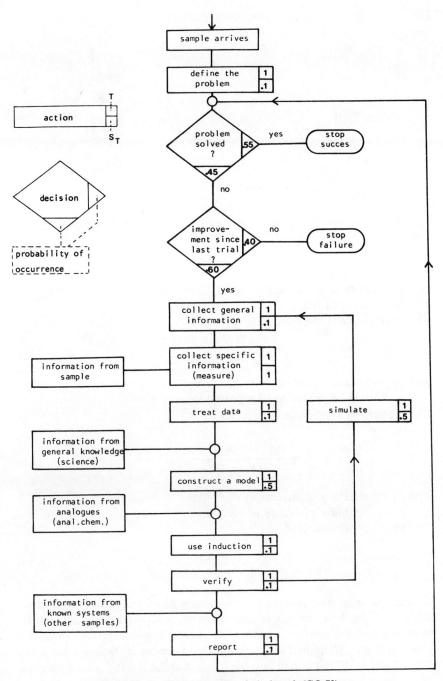

Figure 5.12. Flow sheet of analytical work (GO 72).

247

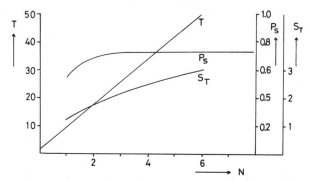

Figure 5.13. Estimation of total time T, uncertainty S_T, and probability of success P_s in the analytical procedure of Figure 5.12.

with t_i = initial time
    t_r = time of repeated steps
    n = number of trials
The uncertainty in the total time is given by

$$S_T = \left(S_{t_i}^2 + n S_{t_r}^2 \right)^{1/2} \tag{5.8}$$

with S_{t_i} = standard deviation of initial time
    S_{t_r} = standard deviation of repeated steps
The probability of success P_S is given by

$$P_S = \frac{p_S \left[\left[(1-p_S)(1-p_F) \right]^n - 1 \right]}{(1-p_S)(1-p_F) - 1} \tag{5.9}$$

with p_S = probability of success in each trial
    p_F = probability of failure in each trial
The ultimate obtainable P_S is

$$P_S_{\substack{n \to \infty}} = \frac{p_S}{1 - (1-p_S)(1-p_F)} \tag{5.10}$$

As can be seen, the probability of success hardly increases after three trials, but the total time increases linearly. The total deviation hardly increases and the relative deviation decreases. The conclusions from this crude and very simplified diagram are that in many instances it probably doesn't pay to do more than two trials and that the result does not increase after three trials no matter how much effort is applied. The time of completion can be estimated more precisely after each trial.

5.2 ORGANIZATION AND MANAGEMENT

At the beginning of this century a number of theories on organization had been developed, based on an approach of the structure of organizational problems. These theories are all focused on efficiency. Taylor and Gilbreth formed a theory that developed into a trend called "scientific management" (TA 11, TA 47). The main characteristic of this theory is that it aims for the foundation of management on conscious, logical, and reasonable principles. Problems with organizations should be approached by empirical methods. Taylor designed a way of organization named "functional organization." In this type of organization one manager guides and controls only one element, for which he has full responsibility.

Another theory is that formulated by Max Weber, who stated that a set of fixed rules should be formulated instead of individual points of view and rules. These rules should be the result of rational considerations. This statement puts the formal properties of a structure in a central position.

The results of a research program concluded by Elton Mayo in the Hawthorne factories of General Electric in Chicago triggered another movement, denoted "human relations." He and his co-workers stressed considerations of the social situation in which the laborers work. His conclusions may be summarized as follows: "It is important to create harmonious relationships within a group of laborers, because this results in a happy group and a happy group is a productive one." In order to create a harmonious group its leaders should be coached and communication programs should be initiated. However, when this theory was tested in practical situations the results were disappointing. It took some time and research before the organization psychologists discovered that needs exist and operate that are even more important than acceptance by the group. Self-development and self-actualization of an individual become more and more important when needs at a lower and more primary level become satisfied. In this respect the work of Maslow (MA 70) should be mentioned.

As a result of a research program conducted by Hofstede (HO 73), it is seen that various factors contributing to a good work situation are rated quite differently by various groups of people working together in one organization. Managers tend to rate the type of work and the natural relations in the company high, whereas unskilled laborers are more interested in job security, safety, mutual relations, and wages. However, the conclusion that the interests of these extreme classes of people in relation to job satisfaction are different is not true. Hofstede considers job satisfaction to be a summation of terms, each pertaining to satisfaction with one particular aspect of the job. An aspect that is rated satisfactory but unimportant in comparison with other aspects does not contribute significantly to satisfaction with the entire job. From this point of view it is seen that factors contributing to the overall job satisfaction are only slightly different when academic staff is compared with unskilled laborers. Each category thinks a challenge to be the most important contribution to personal job satisfaction.

When a hierarchical structure of needs was applied to organizations, two things happened: (1) it gave rise to much criticism of scientific management and the human relations approach and (2) a new development started, called revisionism. Representatives of this trend are Douglas McGregor, Rensis Likert, and Chris Argyris. This trend approaches the organization from quite a different point of view. In particular, Argyris and McGregor noticed that the commonly accepted organizational measures result from a point of view of human behavior that does not correspond with reality. The average human being is, according to this point of view, considered to be lazy, uninterested in his job, and without interest in having responsibility. Such a person wants, above all, a high income and high job security. Argyris and McGregor do not deny that in a particular situation, for example, when classical organization theory is applied, one may act according to this stereotypical behavior. However, based on much research and supported by many investigations, they arrived at the conclusion that the average person likes to work, can be very devoted to his aims, is willing to accept responsibility, and has imagination, ingenuity, and creativity. The organization itself may, in many respects, be the factor responsible for hampering the development of desirable characteristics and the possibility of showing them. McGregor and Likert do not state that in modern business human welfare in general is sacrificed to the highest possible productivity, but that not only human welfare but also obtainable productivity are sacrifices to a fiction (GR 57, GR 60, GR 67).

This new vision of organization was obtained after much research on productivity. The running hypothesis during these investigations was that organizational methods should raise productivity. It appeared afterward that this assumption was not justified. Likert conducted many of these investigations. He analyzed the results of management and staff personnel in business and governmental organizations that obtained the highest productivity, lowest cost, and highest job satisfaction with their personnel. He concluded that they act, in general, according to a pattern of managing that differs from that of staff members with less impressive results. This pattern deviates in important aspects from the one prescribed by the theory of organization in those days. Based upon his investigations Likert could formulate two methods of management: (LI 61, LI 67):

1. *Job Organization System.* Here the rules of the game as set by "classical" organization theory should be followed. The foremost characteristic of the organization is its structure of tasks. People should be selected for their ability to execute these tasks.

2. *Cooperation–Motivation Systems.* Here the principles and methodology of the theory of internal organization have been applied to only a limited extent. A high productivity is aimed for by stimulating enthusiasm of the co-workers. This is obtained by an appeal to noneconomic psychological needs of the team workers. From an organizational point of view this system almost always lacks organization.

According to Likert the most effective manager is characterized by his ability to apply all the techniques of classical organization theory, such as labor study and budgeting.

When a comparison is made between effective and ineffective managers one notices that both types apply the same techniques; however, they do this in quite a different manner. The effective managers offer the results as a tool to the workers to control and organize their labor. The application of the results of these measurements in order to get a firm control on labor is not so much at stake. Discussions and consultations within the team are important in the decision-making process, even when improvement of methods and reorganizations are considered. It is customary that each participant in the team gets all relevant information that the manager himself has got. This information pertains to cost, productivity results of work analysis measurements, and so on.

This attitude does not mean that the management refrains from leadership, but rather that it is operating in a "supportive relationship," which means that the manager supports his team in order to give it the opportunity for getting an opinion. In doing so he assumes—and according to Likert, with reason—that the team members possess adequate knowledge, consideration, and creativity to achieve the ultimate goal. Managers who successfully operate according to this scheme do not rely above all on kindness and amicability in order to get optimal results from their team. On the contrary, they often operate very analytically and critically. However, they know how to create opportunities for their people for optimal self-development.

According to Argyris and McGregor self-development also plays an important role. However, these authors show little appreciation for the possibilities of well-known management techniques. They hardly mention them, but instead stress the unsuppressible need of all adults to behave responsibly. The formal organization does not offer adequate possibilities for this. The concepts of self-development or self-actualization applied by Argyris and McGregor may easily lead to confusion and misunderstanding. Argyris gave an extensive explanation of these properties. In his opinion self-actualization means a realization of the individual talent to achieve in its life-span an increasing amount of activity and independence, interest, pluriformity of actions, time perspective, and self-determination and influence on its surroundings. The individual needs for this differ from one person to another; however, a sane person should always possess them. Argyris and to a certain extent McGregor do not found their opinions on the ethical desirability of allowing people self-actualization. The former states that urge toward self-actualization is an undeniable fact that interacts with the operational behavior of organizations whether we like it or not. The manager has no option to prefer one system of organization over another, he simply has to accept that forces are operating apart from voluntary organizational structures. In the end the shape of an organization and its behavior result from both factors (AR 59, AR 64, AR 71, AR 73).

As may be expected for a branch of science as new as organization theory, the points of view presented here are not final. Recently developments can be noticed that, although they do not seem to be fully crystallized, indicate some trends, pointing toward a synthesis of the technical and human approaches.

In this respect Schein and Bennis (SC 65) stress that a human being is more complex than the rational, economic, social personality seeking self-actualization. Moreover, a human being is not only a complex phenomenon but also very susceptible to changes. A person may develop new motivations on the basis of experience within the organization. Human motivation may be different in various organizations or even in parts of a single organization. A person may be productive in an organization out of different motives. Finally, a person may react differently to a way of managing depending on his own motives, capabilities, and tasks. According to Schein, the relationship between human being and organization is characterized by an interaction and mutual exchange of points of view and negotiations in order to achieve a useful psychological contract.

Another approach starts from the organization itself, seen as a sociotechnical system. This approach stresses the interrelations between technology, surroundings and feelings of the members, and type of organization. Silverman shows that it should be useful to be able to analyze organizations, taking into consideration the ambitions of all its members and their ability to force them onto other people. According to his approach organizations are only the continuously varying product of actions founded on the selfishness of the members.

Still another approach to be mentioned is that of Marck and Simon, who analyze organizational behavior in terms of a general picture of the human organism. They depict an organism that is busy with selecting, deciding, and problem solving but that has a limited ability to perform actions simultaneously and that can pay attention to only a limited amount of information. This approach might prove to be important in the description of a data handling organization such as an analytical laboratory, which contrasts with the behavior of data handling as performed by computers and automatic analysers.

Some investigators (Lewin, Cock, Frenck) describe a social situation as resulting from basic processes or developments. They apply group dynamics in order to obtain social changes, and discuss the operation of two counteracting forces: dynamization, which means development, and structurization, which means fixation. The people that control the (financial) means in general need formalization and structurization, because they focus their attention on a continuous application of the means. In general, managers that operate processes need flexibility and a dynamic attitude because they continuously meet the necessity for adaptation to altering situations. Nowadays most organizations change, voluntarily or not, because of a number of internal and external causes:

- Organizations expand and shrink; they are seldom of stable magnitude.
- Essential standards of society change.
- The influence of the external factors on organization increases.
- The pattern of moral values and standards of members of the organization changes.
- The social factor in organization increases.
- Technical factors in organizations may change dramatically (computer).

In summary, society as well as people change their standards and values, and organizations change within themselves and with respect to their relationships.

One may wonder in what respect analytical laboratories are influenced by the factors mentioned here. It is not feasible to elaborate on all psychological, social, and economic features. Moreover, it is impossible to

select one particular system to operate a laboratory. In general the preference for one system over the others depends on the institute or organization where one operates, and indeed the laboratory may contribute only
partly to the final decision.

5.2.1   Description of an Organization

Various experts on organization theory have tried to give an adequate
definition of organization. There are numerous definitions and a selection
out of the collection of necessity is subjective. The following definitions
should be considered as an illustration of the possibilities focused on
organization as a tool of management. The number of basic principles is
also limited because of the selected definitions. These limitations imply
that the term organization is not used for description of management and
that organization is a tool, not a goal.

- Organization is a complex of relative permanent measures aiming for a
  well operating assembly of means in order to achieve a goal set in
  advance.
- Organization is the ordering of actions and creation of an effective
  assembly of means in order to reach a goal.
- Organization is the creation of effective relations between people and
  means in order to achieve a goal.
- Organization is a social structure, constructed from coordinated positions referring to a common purpose.
- Organization is a distribution of labor among a group of cooperating
  people in order to make the company as good as possible with a
  minimal effort and the highest job satisfaction to all members of the
  group.
- Organization is the complete set of rules and standards designed to
  improve efficiency.
- Organization is a complex system consisting of necessary actions and
  alternative applicable means assembled and governed in order to
  achieve a given purpose. The system of actions and means interrelates
  the cooperation of, for example, human/human, human/group, and
  human/means.
- Organization is a frame of functions, tasks, and processes, embracing
  people that aim for a common goal with limited, alternatively applicable means.

The frame is set by the following:

1. Inventory of functions, processes, and procedures.
2. Creation of bodies to operate in the functions.
3. Effective subdivision of functions and tasks among persons and equipment, including a pattern of relations.

An adequate description for our purpose will contain various features such as the following:

- Functions, tasks, rules, standards, and conditions.
- Processes and procedures.
- Cooperation interrelations, economic interrelations, and so on.

These features may be described from three different points of view, technical, economic, and sociopsychological. Each of these areas has more or less its own stabilized organizational theory.

A technician focuses his attention on the operation, control, and ruling of processes and procedures. The economist formulates the boundary conditions for the technical system. The social psychologist develops a system for the functioning, rule, and harmony of the working community and its relationships. He distinguishes on the one hand a social system for the operation and cooperation of the staff in order to achieve the goal of the company, and on the other hand the private aims of the individual and the team. In such a social system one may distinguish the rules of the game one must obey in order to obtain a well-functioning process. Apart from this, one must stay within the boundary conditions set by economic and technical limitations. An economist considers the ranking and functioning of various alternatives for means of production such as labor, equipment, capital, and chemicals to such an extent that the company goal is obtained with as little effort as possible and/or that the yield is as high as possible and/or that the continuation of the institute (company) is warranted. Here the technical and social systems set the boundary limits. Each of these systems distinguishes conditions, the rules of the game that have to be obeyed in order to allow the functioning. Moreover, each system sets limitations to the other two, which define the boundaries within which each one may operate.

The social system formulates the conditions with respect to human power of endurance—physical as well as psychical—that is, working hours, informal group decisions, and so on. The economic system formulates conditions in minimal yields and budgets, and the technical system gives conditions on instruments, for example. As well as attention, knowledge of the question is involved apart from knowledge of investment and operating costs. The former conditions pertain to the social system, the latter to the economic system. In general, one may conclude that an organization should be considered as a complex system, composed of at least technical, social, and economic subsystems that may be distinguished but cannot be separated from each other.

Now we come to the question of how to describe this system in a scheme that visualizes disturbances, changes, and so on. This scheme should also allow evaluation of the results of actions aiming for optimization. The normal schemes of organization with their treelike constructions are completely inadequate for this purposes; at most they describe a formal part of the personnel organization. The entire picture of the organization is not given. A scheme exists that partly serves these purposes; it is described below.

Figure 5.14 depicts an organization as a system of actions or tasks (i.e., actions not yet performed) and means with a mutual relationship. This scheme is based on the assumption that in each organization a set of fundamental actions and means exists and can be listed. A matrix representation of this scheme gives the means along one axis and the tasks along another. The mutual relationship between a task and the available means are represented by matrix elements. Now the figure summarizes all procedures and processes required for good operational conditions. The system operates well, provided each matrix element contains the right conditions. The conditions pertain to boundary limits and to basic requirements. Extension of the two-dimensional matrix with a third dimension gives an indication of the cost and prices per means per time unit. A technician will preferably focus his attention on the first two axes of the matrix; however, the entire problem covers all axes of Figure 5.14. The social system of the organization is found mainly within the column titled personnel. Here cooperation and harmonious relations between individuals and departments play a role. However, each matrix element may be denoted by incoming and outgoing information streams, pertaining to process and procedures and addresses and where they come from or go to, thus making the social system operational in the entire matrix. The economic system of the organization pertains to all means that carry a price and are consumed on reaching the goal. An important implication of the matrix representation is that each specialist who alters something in his subsystem invokes an action in the subsystems of all colleagues belonging to the organization: The three subsystems mentioned may be distinguished but are inseparable!

Even Figure 5.14 does not quite describe the entire problem, because the organization is a dynamic system of actions and means. Each action

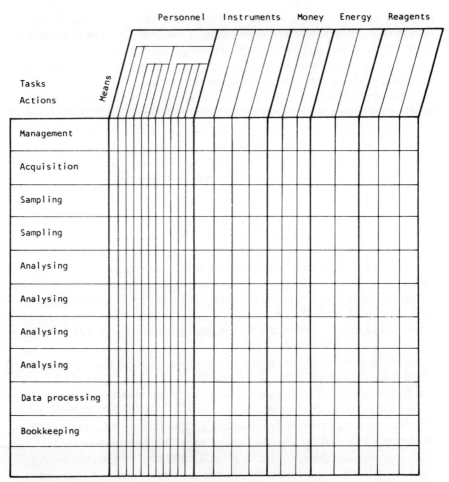

Figure 5.14. Organization scheme interrelating tasks and means as a function of time (SC 66).

requires a period of time for it to become completed. The dimension time is invoked by attributing a time span to each matrix element. Because of internal and external forces the representation matrix of the organization is changing continuously with respect to all dimensions.

5.2.2 Structure of an Organization

The method of representation as presented in the preceding section provides the necessary information on the interactions within an organization but not on its structure as it is or as it should be. Actually it is merely a structure of relationships. Other structures that may be mentioned are a communication structure and a structure of responsibilities or a hierarchical structure. The structure of the communications can be derived from the structures of relationships, because communication means transfer of information. It is possible to visualize the information channels—their location or their desired location—by extracting from the relation scheme the part pertaining to information. One has to keep in mind that information does not pertain exclusively to technical information. Social and economic information should be considered as well.

There are various methods with which to perform communications, the most important ones in a laboratory being the following:

1. *Electric, Pneumatic, or Mechanical Signaling.* Here both emitter and receiver are interconnected with electric, pneumatic, or mechanical channels. A piece of information is coded and emitted and at the receiving end is decoded. The communication network is mostly centralized within instruments, and is unimportant from an organizational point of view. Communication between instruments (e.g., a titrating unit and a computer) as a rule occurs at a level too detailed to be inserted in an organization scheme. At locations where the information stream branches off a further inspection should be made. Here it should be considered whether the branches are optimally positioned or may be omitted (e.g., an electric signal toward a recorder that is not operational).

It is also important to investigate the occurrence of an eventual shortcut in the information stream that may cause a lack of information at a location where it is needed, for example, an operator who does not get the information required for efficient operation of the instruments.

2. *Information Storage.* An information channel that should be mentioned separately is a mechanical one toward paper, recorder strip charts, computer listings, tapes, and paper punchers. The main characteristic of this channel is its ability to bridge time spans. This property has an advantage as well as disadvantage: it builds an archive for seconds up to centuries which may easily be reproduced.

Because of the considerable time delay that may be realized it is sometimes hard to see whether the aim, that is, transfer of information, is reached. The structure of the communication lines, as obtained from the structure of the relations, should be demonstrated to see if the paper communication serves its purpose, taking into account an eventual archivation. Here the questions to be answered are, is the information given to people who should not receive it? Is information filed for future use still up-to-date and useful when required? Is enough information available for interpretation when archives are inspected after a great time delay? Will stored information be inspected in the future?

3. *Acoustic and Optical Communication.* This type of information transfer uses acoustic or optical data channels. In a laboratory the human emitter and receiver of acoustic signals are the most important ones. Optical emitters are found in instruments such as lamps, led displays, and analogue meters such as clocks, and also in paper. Sometimes optical receivers are found inside instruments, such as photodetectors; when they are found outside instruments the human receiver is the most important one. Both types of communication are denoted by high-speed signal transfer, the possibility of handling complicated patterns of information such as pictures, music, and speech and their transient behavior.

4. *Communication Using Human Beings.* It was mentioned above that the human being may act as an emitter as well as a receiver of optical and acoustic information. The human being may also serve as an archive for this information, which enables the bridging of time gaps, as do paper archives. However, important differences to be noted are that human memory is rather unreliable and within the human mind information can easily be confused. These two features are probably interrelated. A resulting advantage is the involuntary and even unknown falsification of information. An advantage of combination of internal communications in the human mind is the relatively easy decoding of complicated patterns of information. A human mind may simplify, interpret, and recognize patterns.

The judgment or construction of a communication structure within a laboratory should proceed according to the following rules:

- Derive the pattern of communication from the information part of the relation structures.
- In situations where a time gap should be bridged, paper or magnetic computer memory devices (tapes, disks, cassettes) should be selected.
- In situations where information should be combined or interrupted or when pattern recognition should occur, human effects should be applied, eventually in combination with computer techniques.
- All cases of straightforward data transfer should be handled along electric, pneumatic, mechanical, optical, or acoustic channels. The last two ways are of special importance for communication with people and for interhuman communication. The other channels are of importance for interdevice communication.
- Communication serves informations transfer; without a receiver a channel is useless!

Finally, a laboratory communication scheme may be presented in Figure 5.15. Here exclusively technical and economic communication lines are drawn. It must be possible to extend this scheme with a social structure. This could reveal some surprising facts. A bad technical communication is often caused by disturbances in the social communication channels: an unapproachable boss, people from hostile communities, or an immigrant worker who does not fully understand the language of the host country are some causes that may easily be detected. Beneficial influences may result from people such as secretaries or assistant managers if they are competent. They make the communication lines shorter and more efficient and operate with less mistakes. Research on this subject revealed that these informal communication lines generally do not correspond with the official lines, which correspond with the responsibility structures (AL 68, AL 70, AL 77, FO 71, FR 71, PE 66).

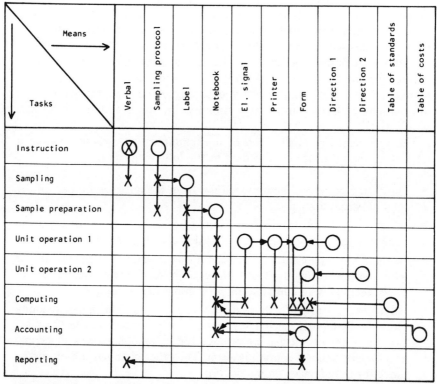

Figure 5.15. Relation between tasks and tools in analytical chemistry; tasks presented in chronological order.

Some other factors to be mentioned follow:

- Geographic distance: there seems to be a reciprocal exponential relationship between communication and distance.
- Social "wavelength" (language, habits, regional origin).
- Ability to interpret (to process difficult information in a manner accessible for the receiver).

5.2.3 Structure of Responsibility

The hierarchical structure of a laboratory depends greatly on that of the entire company. The latter is established by historical, philosophical, and political considerations. Therefore it is not very useful to discuss laboratory organization in general. However, within the responsibility of the department one may have an opportunity of a change in the hierarchical structure, and so some remarks may be made here. In view of the theories of organization discussed above, it seems desirable to delegate responsibilities, as far as possible, to persons or a group of people. The following criteria should apply:

1. The responsibility should be such that it can be taken. The person under consideration should have a personality that can cope with the responsibility in terms of emotion, experience, and knowledge. In other words, one should have the ability to perform all actions that are necessary to do one's part of the job. In order to do this some means pertaining to character, education, budget, and decisive authority are required. One should have the power to apply these means freely, to use them at will and alternatively. A characteristic feature of the situation is that one must account for all duties one is to perform but not for details as to how they are done.

2. Control of the combined duties should be possible. Apart from the means to perform controlling actions one needs information to compare the results of actions with respect to standards. The structure of the organization should reveal whether the person under consideration may get the appropriate information. Important technical information refers to knowledge about the characteristics of procedures that are applied (such as accuracy, speed, correctness) and the standards that should be set (dictated by the process under consideration). Information on economics comprises costs such as cost per analysis, fare composition, overhead, equipment, chemicals, and available budgets. Information on social circumstances

comprises knowledge on merits, labor situations, job satisfaction, disturbances, and so on between interrelated people and people that interact with the group. Standards are found with labor rulings and other formal regulations, social standards, and ethical standards.

The area that is covered by the responsibilities of a laboratory technician is set by a number of features partly belonging to the person and partly to the technician's supervisors. The positioning of the responsibility area within the structure of relationships is governed by various rules. Some demands are as follows:

- The responsibility should be bearable.
- The entire area of relationships should be covered by various responsibility areas.
- The structure of the communication network sets the organization (i.e., for an information-producing organization such as a laboratory) and not the hierarchical structure.
- In situations where an overlap exists, the communication line should point in one particular direction.

Some structures of responsibilities may be presented:

1. *Organization According to Unit Operations.* Each organizational unit, one person or a group, bears the responsibility for one single action or a particular group of actions. The advantages of this type of organization are the minor requirements of knowledge, the ease of automation and mechanization, and the economic application of analysis on a routine base. A disadvantage is the high information transfer from person to person.

2. *Organization According to Procedures.* Here each organizational unit is responsible for one procedure or a particular group of procedures. Qualitative control is of the upmost importance. Its advantage is mainly the minor information transfer between people; its disadvantages are the enhanced possibility of carry over, which here means that analytical results of different samples are interchanged by mistake, and the enhanced requirements of knowledge.

3. *Organization According to Duties.* Each organizational unit is responsible for a particular area of tasks, for instance, the production of information about a particular plant or product or for a research group. The advantages here are a very limited transfer of information between the units and a clear responsibility for production or for the results of research.

The disadvantages are the heavy demand for knowledge and the possibility of a shortcut within the laboratory organization. Some of the communication channels become more or less artificial.

4. *Mixed Organizations.* A situation often encountered in practice, but not tested for optimal conditions with respect to organizational behavior, is as follows: an organizational unit (person or group) is responsible for one task or a combination of tasks and for one procedure or a combination of procedures. Here the entire area is covered better and the lines of communication follow a natural pattern, which is advantageous. The disadvantages are the overlap in the structure and the economic structure which is hard to visualize. It is difficult to denote the costs to particular actions and to calculate the overhead on instruments. An excellent survey of management studies pertaining to analytical research is given by Goulden (GO 74), who describes the consequences resulting from the employment of various managerial styles. Other literature on this subject is listed below.

REFERENCES

AK 76 M. van de Akker, G. Kateman: *Z. Anal. Chemie* **282**, 97 (1976).

AL 68 D. H. Allen: *Oper. Res. Quart.* **19**, 25 (1968).

AL 70 T. J. Allen: *Res. Dev. Manage.* **1**, 14 (1970).

AL 77 T. J. Allen: *Managing the Flow of Technology*, M.I.T. Press, Cambridge, Mass., 1977.

AR 59 C. Argyris: *Manage. Rec.* **12**(3), 82 (1959).

AR 64 C. Argyris: *Integrating the Individual and the Organization*, Wiley, New York, 1964.

AR 71 C. Argyris: *Management and Organization Development*, McGraw-Hill, New York, 1971.

AR 73 C. Argyris: *Harv. Bus. Rev.* **51**(2), 55 (1973).

BO 71 E. de Bono: *Lateral Thinking for Management*, McGraw-Hill, New York, 1971.

BR 75 R. R. Brooks, L. E. Smythe: *Talanta* **22**, 495 (1975).

BR 76 T. Braun: *Talanta* **23**, 743 (1976).

BR 80 T. Braun, E. Bujdoso, W. S. Lyon: *Anal. Chem.* **52**(6), 617A (1980).

DA 63 N. Dalkey, O. Helmer: *Manage. Sci.* **9**, 458 (1963).

DA 70 D. G. S. Davies: *Res. Dev. Manage.* **1**, 22 (1970).

DO 75 K. Doerffel: *Euroanalysis* II (1975).

FO 62 B. Folkertsma: "Mens en organisatie," *Doelmatig Bedrijfsbeheer* **14**(2), 48–50 (1962).

FO 71 J. W. Forrester: *World Dynamics*, Wright Allen Press, 1971.

FR 71 P. A. Frost, R. D. Whitley: *Res. Dev. Manage.* **1**, 71 (1971).

GO 61 W. J. J. Gordon: *Synectics*, Harper and Row, New York, 1961.

GO 72 G. Gottschalk: *Z. Anal. Chem.* **258**, 1 (1972).

GO 74 R. Goulden: *Analyst* **99**, 929 (1974).

GR 57 D. M. McGregor: *Adventure in Thought and Action*, Harvard Business School, Cambridge, Mass., 1957.

GR 60 D. M. McGregor: *The Human Side of Enterprise*, McGraw-Hill, New York, 1960.

GR 62 F. E. Grubbs: *Oper. Res.* **10**, 912 (1962).

GR 67 C. McGregor, W. G. Bennis (Eds.): *The Professional Manager*, McGraw-Hill, New York, 1967.

HA 69 P. D. Hall: *Technological Forecasting and Corporate Strategy*, (G. S. C. Willis, G. Ashton, and B. Taylor, Eds.), Bradford University Press, Bradford, 1969.

HO 73 G. Hofstede: "The importance of being Dutch," Chap. 25 in *Arbeids- en Organisatie psychologie*, (P. J. D. Drenth, P. J. Willems, and C. J. de Wolf, Eds.), Deventer, Kluwer, 1973, pp. 334–349.

KA 64 D. Katz, R. L. Kahn: *The Social Psychology of Organizations*, Wiley, New York, 1964.

KA 67 H. Kahn, A. J. Wiener: *The year 2000*, A Framework for Speculation on the Next Thirty Three Years, *Macmillan, New York*, 1967.

LI 61 R. Likert: *New Patterns of Management*, McGraw-Hill, New York, 1961.

LI 67 R. Likert: *The Human Organization: Its Management and Value*, McGraw-Hill, New York, 1967.

MA 64 J. F. Magee: *Harv. Bus. Rev.* **42**(4), 126 (1964).

MA 70 A. H. Maslow: *Motivation and Personality*, Harper Row, New York, 1970.

MA 75 D. L. Massart, L. Kaufman: *Anal. Chem.* **47**(14), 1244A (1975).

MA 77 D. L. Massart, H. de Clerq, R. Smits: "Reviews on Analytical Chemistry," presented at the *Euroanalysis Conference II*, Masson, Paris, 1977.

MA 78 D. L. Massart, A. Dijkstra, L. Kaufman: *Evaluation and Optimization of Laboratory Methods and Analytical Procedures*, Elsevier, Amsterdam, 1978.

PE 66 D. C. Pelz, F. M. Andrews, *Scientists in Organizations*, Wiley, New York, 1966.

RI 74 T. Rickards: *Problem Solving through Creative Analysis*, Gower Press, Epping, Great Britain, 1974.

SC 65 E. H. Schein, W. G. Bennis: *Personal and Organizational Change through Group Methods*, Wiley, New York, 1965.

SC 66 A. C. Schram: *Tijdschr. Effic. Doc.* **36**(9), 511 (1966).

SC 75 B. Schmidt: *Glas, Instr. Tech.* **20**(11), 1167 (1975).

SC 77 B. Schmidt: *Z. Anal. Chem.* **287**, 157 (1977).

SO 77 W. E. Souder, R. W. Ziegler: *Res. Manage.* **1977**, 34 (July).

TA 11 F. W. Taylor: *Scientific Management*, Harper, New York, 1911.

TA 47 F. W. Taylor: *The Principles of Scientific Management*, Harper and Row, New York, 1947.

VA 74 I. K. Vaananen, S. Kivirikko, J. Koskenniemi, J. Koskimies, A. Relander: *Methods Inf. Med.* **13**(3), 158 (1974)

VA 77 B. G. M. Vandeginste, T. A. H. M. Janse: *Z. Anal. Chem.* **286**, 321 (1977).

VA 79 B. G. M. Vandeginste: *Anal. Chim. Acta*, **112**, 253 (1979).

VA 80 B. G. M. Vandeginste: Thesis, Catholic University of Nijmegen, 1980.

INDEX

Pages in *italics* indicate that the topic appears as a heading or subheading.